Facing Cyber Threats
Head On

Facing Cyber Threats Head On

Protecting Yourself and Your Business

Brian Minick

ROWMAN & LITTLEFIELD
Lanham • Boulder • New York • London

Published by Rowman & Littlefield
A wholly owned subsidiary of The Rowman & Littlefield Publishing Group, Inc.
4501 Forbes Boulevard, Suite 200, Lanham, Maryland 20706
www.rowman.com

Unit A, Whitacre Mews, 26-34 Stannary Street, London SE11 4AB

British Library Cataloguing in Publication Information Available

Library of Congress Cataloging-in-Publication Data Available
ISBN 978-1-4422-6548-6 (cloth : alk. paper)
ISBN 978-1-4422-6549-3 (electronic)

The paper used in this publication meets the minimum requirements of American National Standard for Information Sciences—Permanence of Paper for Printed Library Materials, ANSI/NISO Z39.48-1992.

Printed in the United States of America

To my wife, Anne Lynn:
Thank you for loving and supporting me as I have taken on numerous
"projects." Your trust and confidence in me makes me better. I also want to
thank you for using your literary talents to make this work so much better.
You are a blessing to me.

Contents

Introduction

Like many boys growing up, my friends and I used to play cops and robbers. In that game you are either a "good guy" or a "bad guy," and I always liked being the good guy. There was just something appealing about being the hero who could protect the innocent, stop the bad guys, and generally save the day. While playing cops and robbers, we would invent elaborate and fantastical ways for the good guys to win. They usually involved some level of superhuman strength and agility.

As I got older I came face to face with the fact that I was not the superhero type. In fact, I was more like the ninety-pound weakling. To further drive this point home to my developing mind, about this same time, I discovered that I'm a bit of a computer nerd. My illusions of protecting people and saving the day were soon put behind me. My cape would eventually be stuffed in the back of my closet with my safety blanket.

The upside to this is that I'm one of those people who have always known what they wanted to do when they grew up. I got my first computer in the mid-80s and never looked back. The only question in my mind was what exactly I would end up doing with computers. Whether it would be writing programs for computers, teaching computer classes, or any of a large number of options in between—I wasn't sure. Generally speaking, I liked it all and would have been satisfied with any of them.

I spent a lot of time after college writing computer programs. I loved solving problems for people by making computers do things that others couldn't. I still believe that computer programs are works of art. Not only is there art and elegance in the algorithm, or the way the programmer instructs the computer, but also artful is the code itself and how it is formatted. At one point, I thought about printing some of my code and framing it. Yep, that superhero cape was completely forgotten about and the dorky glasses and pocket protector had officially taken over.

Then, one day, something happened that forced me to dig deep into the closet and pull that cape out. When I was first asked to be a cyber security leader, my primary directive was to keep bad guys from breaking into computers and stealing information. It was like playing cops and robbers. I was the good guy, once again trying to stop the bad guy and save the day, only this time it was for real. To make it even better, the battleground was on computers. I got to put on the cape and keep the pocket protector. What could be better?

As a cyber security leader, I don't really wear a cape, and for the record, I have never worn a pocket protector either. However, myself and many other cyber security leaders like me do view ourselves as defenders of good. In cyber security there really is a bad guy who is trying to break in, steal, harass, and take advantage of people. It is the defender's job to stop these bad guys. I find doing that fascinating.

I personally find cyber security to be one of the most interesting topics. It is not just good versus evil; it is not just technology against technology; it is strategy against strategy. You have to out think the bad guy. In some ways cyber security is like a chess game, and I love this element of the challenge.

As you read this book, we will explore not just the technical aspects of cyber security and how it got to be as large of a challenge as it is today, but we will also explore the personal nature of cyber security and how that needs to drive how we defend ourselves. Cyber security is a rare discipline that combines technology with strategy. I hope you come away from the book as intrigued and interested in cyber security as I am.

Chapter 1

The World Is Introduced
to Cyber Security

By most measures, December 18, 2013, was a normal day. Radio stations had been playing Christmas music for a couple of weeks. Black Friday had come and gone. After all the numbers were in from the big day, retail sales were down slightly, but apparently injuries and deaths were also down with only one death and fifteen injuries reported.[1] Most people in the United States were in the holiday mood and it was still the heart of the holiday shopping season—even if the Black Friday numbers were down slightly, stores and retailers were doing brisk business. Many people were still hunting for a Big Hugs Elmo doll that so many children were asking for.[2] In the middle of this normalcy, a news story broke. Years after this news came out, people do not necessarily remember where they were when they heard it. It was not like the JFK assassination, the space shuttle *Challenger* disaster, or 9/11. However, unlike most news, years after this event, many people still do remember the story. They remember the story as the news that introduced them to a new problem that most had not even thought of before.

A reporter named Brian Krebs revealed that Target stores across the United States were investigating a data breach that may have compromised millions of credit and debit cards.[3] This news was big, but many people were left scratching their heads wondering what a data breach was and why they needed to care. Talking heads—who were "computer experts"—came on the news programs to offer their explanation of the situation and help people understand how this data breach may have impacted them.

For many people, this news story was the first time they had even heard the term "data breach." Many were left wondering what exactly a data breach was. Most people knew what data meant. Most people also had heard the term "breach" used before as well, but put the two together and what does it mean exactly? Was this concept similar to a baby being born breach? Was data coming out of computer systems backward? Or was it like a ship whose hull is breached? Was data flooding into something? Either way, the term "data breach" was related to computers, and many people do not fully understand computers, nor do they particularly care to. The public did realize, though, that this "breach" could not have anything to do with backward data or data floods because something was happening with millions of people's credit and debit card numbers, and that "something" was not at all good.

Most people understand the concept of having their credit or debit card numbers stolen. This can happen if a dishonest waiter wrote down the numbers on his customers' cards or if someone stole your purse and headed to the store. Could this breach, and the many others that have since followed at numerous retailers, have similar consequences? As more and more retail data breaches made the news, it became clear that these data breaches did have similar consequences. In fact, the more these stories developed, the more they became a bit like something out of a James Bond film. Similar to the scene in *Skyfall* where the bad guy Silva hacks into MI6 in order to do all kinds of devious things, including blowing up M's office; to a lot of people, stories behind all the retail breaches seem like something Hollywood made up.

A massive conspiracy was underway to steal millions of card numbers and then sell them to the highest bidders on the black market. The mastermind of this conspiracy would get very rich, and the people who bought these cards on the black market would then use them to buy whatever they wanted. In the old days, something of this scale would have required a criminal mastermind to convince thousands of store employees to covertly write down card numbers and then funnel those card numbers back to a single person. Can you imagine an evil organization like James Bond's nemesis SPECTRE controlling all the cashiers in a large retail chain, with each cashier faithfully following orders from the top of the evil organization as they pilfer card numbers? Unfortunately, this scenario is not very far from what really happened.

The reality of most retail data breaches, is that the criminal mastermind did not have to convince the employees to write down credit card numbers. They only had to instruct the computers in the cash registers to write them down and send them to a central point where the perpetrators could collect them. Because computers don't reason—they only do what they are told to do—there were no moral obligations to overcome. Once the attacker got in a position to send instructions to the cash registers, he was able to create an army of loyal servants that dutifully wrote down every credit and debit card number they saw and send it back to their master.

This is be what essentially happened in all the retail store breaches, however, on December 18, the day the Target story broke, the full scope of that breach was not yet clear. However, as people realized that their credit and debit cards may be at risk, the news spread quickly. The next day, December 19, Target publicly acknowledged the attack, and customers flooded Target's website and customer support lines. For many people, this incident stands out as their introduction to cyber attacks. This introduction was only reinforced as more and more retailers began announcing their own breaches.

Many people learned that a data breach is simply when someone breaks into computer systems and steals data. They also started to question how secure computer systems actually are. For years people thought that as long as they had antivirus software, they would be safe. If for some reason people really wanted to raise the security bar, they would buy something called a "firewall." Antivirus was good enough for most of us, but if you added a firewall, then you were impregnable. After all, hadn't the IT security industry been around for a while and weren't those products the foundation of the industry. Based on this logic, many people were asking how could all these breaches happen? Didn't those companies have antivirus or a firewall? As it turns out, all of them did have antivirus, numerous firewalls and many of the latest and greatest IT security technologies protecting their computers.

Unfortunately, these defenses were not enough. By the end of 2014, Jimmy Johns, Neiman Marcus, Home Depot, Sally Beauty Supply, Michael's, P.F. Chang's, Goodwill, Bebe, K-Mart, Staples, and many other businesses were also breached; and many millions more credit and debit cards were lost. This massive stolen data problem obviously wasn't just an issue with how one company approached security. With so many retailers being breached, this was an issue that ran across the entire retail industry. How could attackers break into secure computer systems and the cash registers that sit behind them in order to change the instructions on those cash registers so that they became part of an evil droid army?

To make matters worse, it was not long after this disturbing news that the public started hearing about potential breaches in the power grid. The grid is made of much more than just the wires that bring electricity to your home or office. The power grid is a very complicated system that manages the supply and flow of electricity, ensuring that the wires going into your home or office have enough juice to power the lights when you flip the switch. Old stories of a computer-based attack destroying a power generator started gaining new life.[4] At the same time, stories of companies losing sensitive corporate data through computer breaches started becoming more prominent. Before long, cyber security was, and still is, on the minds of many people. The public suddenly realized that cyber security is not just a problem in the retail sector, but that it is a problem that all businesses and even individuals need to address.

Not only are credit card numbers that are passing through retail stores at risk, but designs for the next set of technological breakthroughs, research and development intended to take companies into the future, even our personal pictures and many other sensitive pieces of information are all at risk. As we examine cyber security, it is important to remember that the stakes are high.

Today's cyber risks do not just stop when you receive a new credit card. Today's cyber risks threaten the future of many companies and even the economic well-being of many countries.

Imagine what the world would be like today if Silicon Valley was not located in the United States, but in Mexico. Companies like Apple, Hewlett-Packard, Oracle, Facebook, Cisco, and many others would no longer be US companies. For the most recent fiscal year, Apple posted $234 billion in revenue,[5] Hewlett-Packard $52.1 billion, Oracle $37.4 billion,[6] Facebook $17.9 billion,[7] and Cisco $49.6 billion.[8] All the wealth that these companies have created would no longer be part of the US economy. Many of the jobs would no longer be US-based jobs. Now, imagine what this industry would look like if the technology that made these companies had been stolen and used to create different companies in different countries.

Cyber security is not just about credit cards. Cyber security is about the future of many companies and the economies inside of which those companies operate. The stakes are high and the challenges are numerous and constantly changing.

Target was by no means the first shot fired in the cyber realm. However, it is the shot that most people first heard. For many, it is the event that they first remember. That doesn't mean that cyber security is a new challenge though. Cyber security is a growing problem and something that every business leader needs to understand and every individual needs to be aware of. It has the potential to impact the way we do business and live our lives. In the next few chapters, we will take a look at these issues and also examine how we arrived at this point, and what each of us can and should do to protect both our businesses and ourselves.

There are a number of very technical publications on the market that explain cyber attacks and different approaches for defending against them. As we explore these topics, we will make every effort to avoid technical minutiae and present concepts in a way that is clear and understandable to a nontechnical audience. In other words, you do not need to speak geek in order to continue reading. Through a series of examples and illustrations, we will attempt to demystify the tech-heavy cyber security topic, giving you a solid understanding of the challenges in the space, how they have changed over time, as well as what you can do on both personal and professional levels to help protect yourself and your business.

NOTES

1. "Calm Black Friday: Only 1 Death, 15 Injuries Attributed to Big Shopping Day," last modified December 2, 2013, http://business.time.com/2013/12/02/calm-black-friday-only-1-death-15-injuries-attributed-to-big-shopping-day/.

2. "2013's Christmas Toy Hot Lists, Abridged: 7 Things Your Kids Will Want," last modified September 18, 2013, http://www.dailyfinance.com/on/2013-christmas-toy-hot-lists-walmart-kmart-toysrus/#!fullscreen&slide=1008811.

3. "Posts Tagged: Target Data Breach," accessed November 26, 2015, http://krebsonsecurity.com/tag/target-data-breach/page/2/.

4. "Staged Cyber Attack Reveals Vulnerability in Power Grid," accessed November 26, 2015, https://www.youtube.com/watch?v=fJyWngDco3g.

5. "AAPL Key Statistics," accessed March 23, 2016, http://finance.yahoo.com/q/ks?s=AAPL+Key+Statistics.

6. "ORCL Key Statistics," accessed March 23, 2016, http://finance.yahoo.com/q/ks?s=ORCL+Key+Statistics.

7. "FB Key Statistics," accessed March 23, 2016, http://finance.yahoo.com/q/ks?s=FB+Key+Statistics.

8. "CSCO Key Statistics," accessed March 23, 2016, http://finance.yahoo.com/q/ks?s=CSCO+Key+Statistics.

Chapter 2

The Evolution of Security and the Emergence of Cyber Security

Cyber security did not become a problem in 2013 when news of the Target breach first broke. It came into existence soon after computers came into existence. How did cyber security turn into the issue that it is today? How did IT security fail, resulting in the breaches that we all read about? These are the questions we will answer.

In the broadest sense, cyber security is just an extension, or another aspect, of the human race's perpetual struggle to find security. Security has been around since the beginning of the human race. The first humans piled up stones to make a protective wall, and security was born. If you think about it, security has been with us ever since and is really part of who we are.

Over time, that first stonewall grew into a castle, and that castle turned into an underground bunker; all the while the tools used to attack them became increasingly more powerful. As this happened security has evolved and grown more sophisticated. The one constant with security is that it is always changing. As threats change and evolve, the defenses designed to stop those threats must keep up or stay ahead if at all possible. What was secure last year may no longer be secure this year. If a defense does not change, it will eventually become obsolete. This concept is all around us. Whether it is the locks we use for our doors going from simple latches to deadbolts, or the equipment our militaries use, defenses must continually change.

Consider how militaries protect their communication. In wartime especially, it is absolutely imperative that military units protect their internal communications. If these communications are not protected, the adversary can learn about troop placement and movements. They could anticipate attacks and defenses. Letting internal communications leak to an adversary would be a serious disadvantage. To address this problem during World War II, the Germans adopted something called the "Enigma machine" to protect their communications from the Allies.

To protect their communications, Germany used the latest technology in the form of the Enigma machine. This machine was based on an old concept of scrambling and unscrambling messages in such a way that even if someone were to intercept the message, he or she would not be able to make heads or tails of it. At the time, Enigma was a state-of-the-art message defense system. It was very advanced, and when it was first deployed, it was so effective it was considered unbreakable. Using a combination of mechanical rotors and electronic plugs, the Enigma had 150 quintillion (that's 150 with 18 zeros after it) different possible combinations.[1] To make it even harder to break, the code was changed every day. This means that even if the Allies were able to break the code and find the 1 in 150 quintillion combinations that worked, they would only be able to unscramble the message sent during that day. The next day there would be a new code, and the Allies would have another 1 in 150 quintillion chance to figure it out.

In a day before we had computers that could perform millions of calculations a second, trying all 150 quintillion combinations to find the one being used that day was virtually impossible. No wonder the Germans were confident in the security of their communications! They relied on the best technology at the time and mathematically speaking, the code was unbreakable. Fortunately for the Allies, neither the technology nor the odds deterred them from trying to break the code.

The British set up a dedicated site at Bletchley Park to focus on code breaking during the war. Staffing this effort was one of their highest priorities. When asked to obtain more people to help with the code-breaking effort, Prime Minister Winston Churchill responded, "Make sure they have all they want on extreme priority and report to me that this has been done."[2] At its peak, 9,000 people were working at Bletchley Park.[3] This staff worked six days a week on eight-hour shifts with only half-hour meal breaks. With only a one-week leave four times a year, the environment was very demanding.[4]

Through some of this work, engineers developed a device known as a "bombe." This "bombe" was a hybrid electronic and mechanical device that helped code breakers figure out what settings the Germans were using on their Enigma machines. These settings determined which of the 150 quintillion possible combinations was being used. If the Allies could determine the settings, they could break the code and read the German communications. Each bombe machine was seven feet high and seven feet wide. They were two feet deep and weighed right around a ton.[5] These machines had the ability to churn through all the 150 quintillion possible code combinations to identify the combination of the day. At their peak, these machines allowed Allied intelligence to read around 4,000 messages a day.[6]

With enough time and resources, the people at Bletchley Park had figured out how to overcome some of the best message security of their day. This research was a turning point in the war. The Supreme Allied Commander, Dwight D. Eisenhower, considered the breaking of the code to be "decisive" for the Allies eventual victory.[7]

This "decisive" turning point is just one example of how security constantly evolves. It reminds us that what is secure today will become insecure tomorrow. Even items that we use every day, such as locks, need to change in order to keep up with new threats. In 1851 Linus Yale invented the "Magic Infallible Bank Lock."[8] Linus went on to found the Yale Lock Company, which many of us may recognize today, but someone found a way around this sophisticated, magic lock, and even though you can still purchase Yale locks, the infallible model is no longer on the market today.

Regardless of whether you are talking about locks, protecting communications, or any other security-related realm, security is inherently an arms race between attackers and defenders. The moment one raises the bar, the other must respond. We can never do enough for security's sake. Most of us can see and understand this evolution when it comes to weapons, fortifications, or even locks. The spear gave way to the bow and arrow, which gave way to the gun. These changes happened over the course of centuries. Because of this pace of change and that it took lifetimes for these technologies to change, people did not necessarily feel as though a revolution had taken place when they did change. The change felt more natural. It made sense.

Enter the computer age: a time when things change at breakneck speed. What was new last year is old this year. Just think about how long you have your cell phone before it starts to feel old. The speed and scale of change are happening at unprecedented rates and regardless of who you are, it is difficult to keep up. Insert the ever-changing nature of security into this technical revolution, and the speed of change that it enables, and you can easily see why cyber security is a complex and confusing topic. It is changing so fast that it is hard to get your hands around it.

The ever-evolving and advancing nature of security is what makes cyber security so scary—and yet so interesting—to those who live it every day. I've protected national secrets. I've protected some of the largest companies in the world. I've been part of the fight to keep the bad guys at bay, but I'll share a little secret with you: I never wanted to be an IT security professional. In fact, the thought of IT security professionals used to conjure up images of people who worked in the bowels of computer centers—constantly mumbling to themselves about vulnerabilities, audits, or policies. In my mind, they were isolated. That is, of course, unless they were telling someone "no." Let's face it, who wants to be the naysayer in an industry powered by innovation? No, I never aspired to be an IT security professional.

But all of that changed. Just like the Allies' ability to read German communications during World War II, whether anyone realized it or not, something big changed, something "decisive" happened in IT security. No one announced the change on the news. No trade magazines published it. In fact, most security professionals today still don't know that it happened. Yet this change, or shift, turned me into an IT security professional. This very change got so many big businesses in the news as the victim of a breach. This same change will drive how businesses must look at security. This shift impacts so much that it will change a large part of how businesses think. It is because the change is so significant that I use the term "IT security" to refer to what people did before the change, and the term "cyber security" to refer to what people must do after the change. Let me explain.

To understand how IT and cyber security got to where they are today, we need to look at how IT security evolved, where it came from, where it is today, and how that has left the door open to attacks and the need to shift toward cyber security. Historically, IT security was a subculture inside the already geeky IT community. Just like a good police officer who knows the ins and outs of his or her beat, who knows where everything should be, and the routines of those who live along it, IT security professionals knew the ins and outs of IT systems. They knew the details of how these systems worked, what these systems required to work, and where everything should be. A lighter side also existed to what they did. I can remember one group of security professionals who figured out how to play any sound on any computer they wanted. They debuted this newfound capability by playing fart noises on every machine their boss walked past. Ah! The good old days.

If you asked them, these uber geeks would tell you that they stopped bad guys, or hackers, from causing trouble. That's what they thought. That's what they said. There was a sense of pride around being one of the good guys. However, if you dug deeper, you found that in reality they were not stopping bad guys, but they stopped the computer programs the bad guys were writing. It was almost like the United States and Russia fighting proxy wars. IT security professionals were not addressing the people problem, but the technology problem. This difference is subtle but important.

These bad guys, or hackers, wrote programs either for fun, prestige, or just to demonstrate their technical prowess, and then unleashed the programs on the world. Like your favorite musician spending months creating a new album and then releasing it, hoping it will be make a big impression, so did the hackers spend time making their malicious programs hoping they would make a big impact. The hackers' main goal was to create programs that spread on their own, infecting as many computers as possible and causing as much chaos as possible. If the creators of these destructive programs succeeded and caused enough trouble, they could get on the news, and for a lot of them, that was pretty cool. An IT security professional's main job was to keep these malicious programs, also called viruses and worms, and other forms of malicious software from wreaking havoc on computers.

While this game of cat and mouse between hackers and IT security professionals was being played out, this thing called the "Internet" burst onto the scene. As the Internet grew and everyone clamored to get on it, suddenly every computer connected to the Internet could communicate with every other computer on the Internet. This was great for sharing data, but as so often happens, what was invented for good purposes eventually gets exploited for more nefarious uses. In this case, hackers figured out how to use the Internet to spread their malicious programs.

Up to this point, the hackers' biggest challenge was figuring out how to spread their malicious software from one computer to another. If you think about an infectious disease spreading from person to person inside a community, that disease cannot spread if people do not interact with each other in some way. The same concept applies with malicious computer programs. If computers do not interact with each other in some way, there is no way to spread the malicious program. Early on, this was the biggest challenge that hackers had to overcome. Before the Internet, most computers kept to themselves. They really didn't interact with other computers very often, or if they did, they only interacted with a small group of other isolated computers. This meant that the spreading of malicious programs from one computer to another was very challenging.

The Internet completely changed the game. As computers began to access the Internet, they became exposed to a much larger community of other computers. This exposure gave hackers a way to automatically spread their malicious programs from one computer to another. Hackers finally had a way to communicate with almost every computer out there. Rather than sharing information, they shared their malicious software, whether the recipient wanted it or not. This was when hackers really came into their own. Their programs spread far and wide and some of them even made the evening news! Their programs spread quickly, easily, and prolifically, and they finally did some real damage. Yes, the hackers thought that was extremely cool.

As you can imagine, this was a big move and the IT security community had to react. However, rather than focusing on the people who launched these attacks, the security community continued to focus on the malicious programs themselves. True to their background, IT professionals developed technology to address the growing problem. They put software in place to quickly identify these malicious programs, figure out how to break them, and then push that breaking capability to as many computers as possible. This process is industrialized computer security on a massive scale. You may recognize this process as what we know as antivirus software.

With the antivirus approach, we assume that hackers will write one bad program that tries to infect many computers; after all, this is what pretty much every malicious program to this point had done. The goal of the malicious program was to infect as many machines as possible, so building a solution that assumed this behavior would continue just made sense. Based on this assumption, the goal of the antivirus software is not to stop the program from causing trouble on the first few machines that it infects. The goal is to just stop the malicious program from causing wide-scale trouble across many machines. In other areas, this kind of approach would be called a "safety in numbers" approach. A couple of members of the herd will get picked off when the attack first starts, but the antivirus software will ultimately protect a vast majority of the herd.

People played the game this way for decades. Hackers would make new malicious programs, security professionals would find them, figure out how to break them, and then use the antivirus software to deploy that breaking capability to a massive number of computers. During this time, most computer users were relatively safe. The average computer users were aware that computer viruses existed, but the viruses didn't really bother people on a day-to-day basis. This safety-in-numbers approach—based on the assumption that the malicious programs will try to infect as many computers as possible—was paying off. Occasionally a new malicious program would break out and spread to a large number of computers before anyone discovered it, but still, most computer users were safe.

This approach was extremely effective, and we have all benefited from it at some point. Our antivirus program would open a small window saying that it stopped something, we would feel good, and then we would move on with life. All was right with the world.

Building on this success, IT security professionals started to assemble lists of best practices that represented ideal protocols for maintaining strong defenses and responding to issues. These lists included things like making sure you ran the most current version of software packages or that you turned off people's accounts when they left your company or even offering training to your employees reminding them about good security practices, like not writing your password on a sticky note and placing it in your desk drawer (not that anyone has ever done that). These actions were all good to do, and doing them helped to make sure bad things didn't happen on computers. However, it didn't take long for these lists to turn into official "frameworks" that could be audited against.

Once these lists of best practices where audited, they moved from being something that your security leader peers recommend to being a measuring stick with which to gauge the effectiveness of your security program. If a specific list represented all the best things you should do to make your computers safe, then it begs the question: How many things on the list does my security program do? Entire industries soon cropped up to answer this pressing question. IT security was no longer just about providing a strong antivirus solution, but it now helped you implement these frameworks and audited how well you adhere to them. The concept is if you do all the best practices on the list, you'll be secure. As the industry made this shift, the goals of many security teams became tied to passing their audits. Passing audits became synonymous with good security. Never mind the fact that a number of IT security teams were no longer watching whether they were effectively stopping bad guys or not. Bad things only happen when you don't follow the list, right?

Based on this approach, IT security teams today spend a lot of time and resources making sure they do everything on the list. Take, for example, something called "vulnerability management." A "vulnerability" is a flaw in a computer program that a hacker can exploit to do something detrimental. These vulnerabilities sound pretty serious, and some of them are. However, there is a nearly endless list of these vulnerabilities and that list grows daily. In 2013, for example, there were 10,472 new vulnerabilities disclosed[9] by a popular vulnerability-tracking organization. That averages to over twenty-eight new vulnerabilities a day. Based on the sheer volume of vulnerabilities and the large amount of effort required to close some of them, you can understand the enormity of trying to manage all vulnerabilities.

An entire industry is built around managing vulnerabilities inside IT environments. This is called "vulnerability management." Identifying these flaws, prioritizing them, and correcting them can consume numerous resources. As these new security flaws are discovered on a regular basis, finding and fixing them is a never-ending task.

Managing access to computer systems is also a never-ending task. People come and go, and the data they need for their jobs changes regularly. Tracking who can access what systems and whether they still need access to these systems is another time-consuming job. Entire companies have been created to service a market segment known as "access management." Companies spend millions of dollars on software, tools, and audits to help them manage access to computer systems.

These examples, vulnerability and access management, are just a couple of practices that are on the best-practice lists. As you can see, these lists get fairly complex very quickly. They require a lot of time and attention. In fact, there are IT security professionals who focus on just vulnerability management or access management. You can have an entire career dedicated to just one of those disciplines.

IT security is currently focused on these best-practice lists, making sure everything on the list is addressed in a consistent and repeatable way.

Select your list of best practices, do what is on your list, and congratulate yourself when an auditor verifies you are doing everything on your list. The work is tedious, but if your auditor is happy and gives you a passing grade, you at least know you are doing the right things. Or are you?

This model is structured, well researched, and mature. Industries with multiple career paths exist inside it. Many large, well-known businesses focus on it. In fact, one of the largest IT security companies uses a checkmark as their logo. It's almost as if they are implying: use our products and you will get the checkmark you want from your auditor. But have we become so myopic on our lists that we've missed the point? We understand how we got here, but we need to ask ourselves: Are we missing the forest for the trees? Has IT security gotten so consumed by details that it's forgotten what it was trying to accomplish in the first place?

To bring this short history of IT security full circle, notice that IT security started off attempting to stop bad people from doing bad things. Instead of focusing on stopping people, it now focuses on doing what's on a list. Rather than defining success as stopping a bad guy, it defines success as passing an audit. Add to this the mainstay of the IT security industry, antivirus, is based on an assumption that malicious programs will try to infect as many machines as possible, and you have a concoction for trouble.

The change that brought me into the security industry happened within the context of this mindset. The change I am talking about is the rise of targeted attacks.

With defenders focused on their lists and the antivirus industry focused on stopping widespread infection of bad programs instead of hackers, the time was perfect for hackers to start developing destructive programs that were designed to attack one system or a small number of systems. When you think about it, this approach is amazing. Hackers understand what's on everyone's checklist, so they can avoid being caught by security's best practices. They also understand that antivirus is based on the safety-in-numbers assumption. By targeting a handful of computers, they can fly under the radar of the antivirus software. Having their malicious program on their victim's computer, they steal passwords and then use them to log in and masquerade as actual employees. This serves to evade all other defenses. This makes the attacker go virtually undetected. As we have looked at the history of IT security, you can see that this strategy that attackers adopted is honed to hit IT security systems in their blind spots.

This change is seismic. This change turns IT security on its head. Suddenly, the tactics that worked so well for so long are actually being used against us. Suddenly, the shift from focusing on stopping bad guys to stopping bad programs becomes relevant. Suddenly, running your defenses off of an industry-standard checklist, ensuring your defenses looks like everyone else's, becomes a liability. Suddenly, assuming all malicious programs will attempt to infect as many computers as possible becomes invalid. Congratulations, you complied with the payment card industry (PCI) security checklist. This list has all the best practices for keeping credit card numbers safe. It is taken so seriously that computer systems cannot process credit cards unless they are audited and found to comply. You've successfully done all these best practices as defined by the experts who know the most in the payment card industry. You are what auditors call "PCI compliant"! Spoiler alert! So were most businesses that lost its customers' credit card numbers.

I was first introduced to IT security when this seismic change happened. This industry shift—from stopping programs and managing checklists to stopping bad guys through cyber security—brought me into the space. Cyber security returns IT security to its roots: stop the bad guys. In the past, focusing on bad programs was effective, but as hackers customize attacks on targeted victims, security professionals must augment their focus to become aware of the hackers themselves.

On the surface, this change may seem small, but it's not. The implications of this change reach very far. Let's think about it in terms of playing chess. In many ways, security is like a chess game. Two sides try to outplay each other. The attackers try to accomplish their goals, and the defenders try to stop them.

Think about what kind of strategy you would use if you were playing chess against a computer. Computers are powerful because given the same input, they will generate the exact same output every time at lightning speed. Computers are known for reliable calculations, not so much for creative thinking. They also tend to be predictable. Given the same input, they will generate the same output. Therefore, the hardest part about beating a computer at chess is figuring out the right set of moves that win. Once you figure out the moves, you can win every time. Once you determine inputs (moves) that give you the desired output (a win), you simply have to provide that same input over and over and you win over and over.

Growing up, my first computer was a Commodore 64. I had a game named Chessmaster 2000 on it. After much trial and error, I figured out a handful of moves that would beat the Chessmaster. Oddly enough, every time I used those moves, I won. Anytime I wanted to feel good about myself, I could whip out those magic moves and prove that I'm better than the computer at chess. It was fantastic.

Historic IT security is very similar to my Chessmaster game. To stop the bad program, defenders have to figure out one set of moves that will break it. Once they figure that out, they can use those moves over and over to win every time. The checklists, and ultimately IT security, follow this input and output logic.

Now, consider playing chess against a person. Unlike computers, people are not good at doing the exact same thing over and over. People are better known for creative thinking. In fact, their ability to do calculations is often flawed, something I am constantly reminded as I check my daughter's math homework. Based on this concept, if you were to stop playing the Chessmaster and sit down with another person, how would your experience change? Let's assume you won your first game and then just like you did with the computer, you played that same game a second time. What would the outcome be? I wouldn't imagine that kind of strategy would go very well for you. When you play chess against people, you need to understand how they play the game, understand their strengths and weaknesses, and then adjust your game to match your strengths to their weaknesses. Every game you play is different. This is what cyber security is about and why the historic approach of IT security no longer works. After the change, we are no longer stopping programs. We are stopping people.

Stopping people is not about following a checklist. Stopping people is not about hiding in the herd. It is about understanding your opponents' strengths and weaknesses, as well as your own strengths and weaknesses. Understanding that the attacks we hear about on the news and read about in the paper are a fundamentally different problem than what the traditional IT security industry has focused on is the first step in understanding how to address them.

As subtle as the differences between the old IT security approach and the new cyber security approach may seem, they do have deep implications in how companies and their leaders need to look at their businesses. We will explore these implications in future chapters. Before we do that, let's lay some additional groundwork by looking at this thing we call the Internet and how it has gotten us into these cyber challenges.

NOTES

1. "Cracking the Enigma Code," accessed November 26, 2015, http://www.nationalarchives.gov.uk/spies/ciphers/enigma/en1.htm.

2. Copeland, Jack (2004). Introduction to chapter 7. In *The Essential Turing: Seminal Writings in Computing, Logic, Philosophy, Artificial Intelligence, and Artificial Life plus The Secrets of Enigma*, by Alan Turing (ed. B. Jack Copeland), p. 336. New York: Oxford University Press.

3. Copeland, Jack (2004). "Enigma." In *The Essential Turing: Seminal Writings in Computing, Logic, Philosophy, Artificial Intelligence, and Artificial Life plus The Secrets of Enigma*, by Alan Turing (ed. B. Jack Copeland), pp. 217–264. New York: Oxford University Press.

4. "Bletchley Park," accessed November 27, 2015, https://en.wikipedia.org/wiki/Bletchley_Park.

5. "2. Description of the Bombe," accessed May 1, 2010, http://www.ellsbury.com/bombe2.htm.

6. "The Turing Bombe," accessed November 27, 2015, http://www.rutherford-journal.org/article030108.html.

7. Winterbotham, F. W. (2000) [1974]. *The Ultra Secret: The Inside Story of Operation Ultra, Bletchley Park and Enigma*, London: Orion Books Ltd, p. 16, 17.

8. "Linus Yale Jr.," accessed November 28, 2015, https://en.wikipedia.org/wiki/Linus_Yale,_Jr.

9. "Category Archives: Vulnerability Statistics," accessed November 27, 2015, http://blog.osvdb.org/category/vulnerability-statistics/.

Chapter 3

How We Got Here—A Brief History of the Internet

The Internet is not new, but to many people, cyber crime *is* new. Has the Internet changed to allow cyber crime to become the issue it is today? Why does cyber security seem to be an issue all of a sudden? The answer to these questions lies in both how the Internet connects the world and the history of how the Internet was created and became the one central communication platform that is able to connect the world. If you want to understand the origins of recent cyber security issues, you have to go back to the origins of the Internet and to the power of the connectivity that the Internet provides.

Before the rise of the Internet, cyber security was not a big problem. Just like diseases running through a community, it isn't until people start to interact with each other that a disease is able to spread. Similarly, it wasn't until computers could interact with each other that cyber security became a problem. With the advent of the Internet, for the first time in human history, we all have the ability to communicate to a very large part of the world. This communication ability is amazing. Everyone on the Internet can communicate with anyone else on the Internet. This ability is very powerful and will continue to change the way we learn, shop, play, and relate to each other. In short, communicating with the world via the Internet is changing the way we live.

Consider what happened in Tunisia on December 18, 2010. On that day, a revolution began and spread throughout other countries in the Arab League. This revolution consisted of a series of demonstrations and protests that became known as the "Arab Spring." By early 2012, rulers from four countries were forced out of power, and the political face of the region had changed. What is striking about these events is not just the scope of them, but how social media, built on top of the Internet, played such a large role in them. Social media was such a prominent part of these revolutions that they are also referred to as the "Twitter Revolutions." Using this newfound capability that the Internet provides to communicate to a very broad audience, organizers used social media to both coordinate participants in the demonstrations and to shape political debates by spreading awareness of other events around the world.[1]

This power is a double-edged sword. Not only can you communicate with the world, but the world can also communicate with you. The vast majority of people on the Internet want to use this communication capability for good, but there are some people who seek to exploit it.

In the physical world, we understand the value of living in a good neighborhood. People will pay more to live in a low-crime area and to distance themselves from people who commit crimes. This is illustrated by a study conducted by Leigh Linden and Jonah E. Rockoff on the impact of crime on property values. Linden and Rockoff learned through the study that houses sell for 19 percent less when they are located near a registered sex offender.[2]

We count on the people who are committing crimes in other areas to stay in those areas. For the most part, they do stay away. But, what would happen if every criminal in the world could make their way to your front door in an instant and with no effort on their part? Would your safe community continue to be as safe as it is? Delivering instant access to your community makes it just as convenient to commit a crime in your community as it does to commit a crime in any other neighborhood. Instant access removes the physical barrier. Removing the physical barrier is exactly what the Internet has done. Not only can you communicate with anyone in the world, but also every criminal can now communicate with you.

The Internet is inherently a bad neighborhood, providing every criminal the ability to instantly communicate with you. I often find it amazing when security leaders talk about the thousands of "attacks" they stop every day. At first blush, these numbers sound very impressive, and I think security leaders use these numbers to prove how good they are at finding or preventing attacks. However, what we need to look at it how these security leaders actually define an "attack"?

Again, the Internet is a bad neighborhood. But even within that neighborhood, there are different levels of risk. In the Internet neighborhood, an unknown number of people do the digital equivalent of turning doorknobs. They go around every computer they can find to see if all of their proverbial doors are locked. If the doors are locked, they move on to the next computer. These types of "attacks" happen thousands of times a day. These "attacks" are the kind of attacks that security leaders are talking about when they discuss the thousands of things they have stopped. Don't get me wrong; there is value in stopping these low-level attacks, but these doorknob turners are not the only threats we face.

Another group of attackers is out there. These attackers are not just looking for an unlocked door of which they can take advantage. In the cyber realm, criminals exist who really like *your* TV and will stop at nothing to get it. This means if your door is locked, they will try your windows; if your windows are locked, they will break one; if you put bars on the windows, they will cut a hole in the roof. In other words, these people are not randomly attacking you, they are actively targeting you. As you can imagine, these attacks do not happen as often, but when they do, they are much harder to defend against.

The next time your security leader tells you about the thousands of attacks, he stops every day, ask him how many of those attacks were random attacks that just happened to try the locks on your doors compared to truly targeted attacks that are focused on your organization. A big distinction exists, and defending yourself against one type of attack is very different from defending yourself from the other type. We will explore more about these differences later in this book.

Whether you are talking about doorknob turners or targeted attacks, either way, the Internet is inherently a bad neighborhood. This begs the question: How did the Internet become a bad neighborhood? How did a technology intended to bring the world together to freely distribute information and improve our lives turn into something that could be used to attack us? The answer lies in the creation of the Internet: what it was originally intended to do, how it was built, and how it grew.

The origins of the Internet can be traced back to 1969 with a network known as ARPANET. This network allowed computers at four universities to communicate and share information. These universities were the University of California, Los Angeles; Stanford Research Institute; the University of California, Santa Barbara; and the University of Utah. To be clear, ARPANET was not created to change the way the world communicated. The vision for it was not to be something that would stretch across the world and communicate with every computer, smart phone, and refrigerator in the land. As is often the case, big things often have modest beginnings. No, ARPANET was started to simply help researchers share information and work together.

In the 1960s the Defense Department's Advanced Research Projects Agency (ARPA), now known as DARPA, was sponsoring a number of research projects. People working on these research projects were all using various computer systems. Over time, the researchers realized they could more easily share information and work more efficiently if their individual computers could communicate with each other. At the time, each project had its own computer that the researchers on that project shared. When a researcher was working multiple projects, he or she would have access to multiple computers. Different researchers working on various projects shared these computers and depending on who researchers wanted to share information with, they found themselves moving from one computer to another.

Bob Taylor was one of these researchers and a pioneer of the Internet. He recalls how he used to connect to three different computers from his desk. After a while, this became tedious, and he saw the need for his computers to be able to communicate with each other, so he could just get on one computer and share information with anyone he wanted.[3] Taylor said, "Oh man! It's obvious what to do: if you have these three [computers], there ought to be one [computer] that goes anywhere you want to go. That idea is the ARPANET."[4] That's right, the roots of the Internet do not stretch back to a master plan for connecting a large part of the world's population. The roots of the Internet stretch back to allowing a researcher to go from having to jump between three different computers to being able to do everything he wanted on just one. It was about making a small group of researchers' lives simpler, enabling them to more easily share information and work much more efficiently. It just so happens that they solved those problems so well that most of the world wanted a piece of it. What started as a small solution to help a couple people has grown into the worldwide phenomena that we know as the Internet.

With these modest beginnings, we can start to see the seeds of our current cyber security issues. After all, when you are making something to be used by you and your friends, by you and people you know and trust, how much thought do you actually give to security? When you trust everyone, chances are, you are going to spend more time finding a solution to your problem and very little time making sure that solution is secure.

Add to this inherent trust that existed between colleagues the fact that the US government was sponsoring the work, and you have yet another set of drivers to consider. As this network grew, more and more people wanted to connect additional computer systems in various different ways. Originally, all the computers were connected via the equivalent of telephone lines. As the network grew, people wanted to start using things like radios and satellites. This presented a challenge.

With all of these various types of computers all speaking their own language and connecting in different ways, how does a computer know what language to speak and which connection method to use when it tries to communicate with another computer? Think about it this way: Calling your best friends on the phone is an entirely different activity than reaching them on a shortwave radio. If you are going to call on a phone, you can simply pick up the phone, dial their number, and wait for them to answer. Even if they are not expecting your call, their phone will ring, alerting them to the fact that you are calling. Compare this to communicating via shortwave radio. Radio does not have a ringer. If you are going to communicate via radio, you will have to prearrange a time when your friend will be listening for your message or keep sending your message over and over until your friend eventually turns on her radio, hears your message, and responds. The logistics needed to communicate via these two different methods is very different. Add to this the complexity that your friends speak English, French, and Spanish, and you're not sure which language they will be using when they receive your message.

Computers faced these kinds of challenges. Each time a computer wanted to connect with another computer, it had to figure out first how to connect with it and then what language to speak. With many computers only able to connect via one or two mechanism and only able to speak one or two languages, there were a lot of problems and limitations.

The solution to these challenges was what is called a "protocol," in this case, a set of rules that define how computers should communicate with each other. The protocol that was developed is known as the Transmission Control Protocol and the Internet Protocol, or TCP/IP.

In the 1970s, Robert Kahn and Vinton Cerf joined forces to create what eventually became TCP/IP.[5] Cerf recalls, as they developed their solution, "We were well aware of the importance of security."[6] As they laid out their design, they considered adding something called "encryption" into the design. This decision was a pivotal moment for the future of the Internet. If encryption was included in the design for TCP/IP, then messages between two computers would be scrambled as the messages traveled across the network. Just like the German codes in World War II were scrambled when they left the sender and then unscrambled by the receiver, if encryption was designed into TCP/IP, the privacy of messages sent across the Internet would be much better protected. Without encryption, anyone who sees the message as it travels from one computer to another could read it. Additionally, encryption would also make it much easier to verify the source of a message and to make sure you are actually communicating with the intended computer.

As they designed TCP/IP, Khan and Cerf were at a major crossroad for the Internet's future security. Incorporating encryption into the design would pave the way for a more secure path for the Internet. Leaving encryption out would send security of the Internet down a different path. Even with all the benefits of encryption and the inventors stating they were very aware of the importance of security, encryption was ultimately left out of the design. The opportunity was missed and as a result, one of the underlying protocols that make the Internet work moved forward without the benefit of these security capabilities.

How did this happen? The designers knew about the need. The technology to solve the problem was there. Why was encryption not designed into the way computers communicate with each other? As it turns out, a number of good reasons exist. One of these reasons was due to the influence of the National Security Agency. You see, when your job is to read the messages of your adversary, like Alan Turing did for the British during World War II, you really don't want every message protected with sophisticated encryption. In fact, encryption techniques were closely tied to national security and subject to the same kind of export restrictions as military technology. When you think about it, this made a lot of sense. The German Enigma Code machine was essentially an encryption technology. During the war, Germany certainly wasn't shipping that technology to the Allies. Why would a project being funded by the Pentagon allow similar technology to be included in something that was going to be used for nonmilitary purposes?

Ultimately, the National Security Agency was very interested in using encryption to protect military communications. However, using it to protect nonmilitary communications not only made their job harder as eavesdroppers, but it also tipped the government's hand as to how it protected its own communications, potentially making them easier to crack by other governments. From this perspective, we can understand some of the pressure that Khan and Cerf were experiencing.[7] Add to this the fact that, at the time, encryption still had sizable technical challenges. As it turns out, even with computers, encryption is not easy. It required a large amount of computing horsepower to make it work. There were also complexities in managing the secret decoder rings, also known as "keys," which allowed messages to be scrambled and unscrambled. When we consider these challenges, along with the National Security Agency's concerns, we can understand how encryption did not make its way into TCP/IP.

With the creation of TCP/IP, one of the first turning points in the history of Internet security was passed, but it was not the last turning point. It was not the last opportunity to build security in. As the Internet grew and more and more computers became a part of it, new challenges presented themselves, and with these challenges came additional opportunities to insert security. As is often the case when making something new, the first goal is to just make it work. Once it works, refinements can be made from there.

This approach is true of most inventions and applies to the Internet just as much as anything else. We can illustrate this approach by telling the story of how another of the foundational technologies of the Internet was developed. As engineers set out to expand the usage of the Internet and create what has become the Internet that we know today, one of the first challenges they had to overcome was to create a way for two computers to find and connect to each other. Where TCP/IP made sure the computers spoke the same language and were able to use different means of connecting (phone lines, satellites, radio, etc.), there was still a need to define how computers found each other to even have a conversation. In other words, the Internet had a standard language that was spoken, and each computer had proverbial phones to call each other with, but they did not have a phone book to know which number to call.

By 1989, the idea of computers being able to communicate with each other and share information was catching on, and the Internet was growing very rapidly. So rapidly, in fact, that it was starting to hit limits in its ability to grow.

At its heart, the Internet is a massive web of connected computers. For any one computer to communicate with another, there are many, many potential paths through that web that the communication can take. With no definitive map and no central authority to manage this web and the communication on it, something had to be done to allow computers to successfully navigate the web as it grew and became increasingly complex.

As more and more computers joined the Internet, it became clear that the Internet was quickly heading toward a cliff. At some point, one too many computers would join, the web would become too complex, and computers would not be able to find each other. To address this problem and to make sure the Internet didn't melt down, two engineers, Yakov Rekhter and Kirk Lougheed, met for lunch and started drawing on napkins. Three napkins later, the Border Gateway Protocol (BGP) was born. Affectionately referred to as "The Three Napkin Solution," BGP was viewed by its inventors as a short-term stop gap to just keep the Internet going as it grew larger and larger. BGP was something that would get these engineers over the immediate bump until they could come up with something better. They fully expected this solution to be replaced with something more robust. Surprisingly, as of this writing, over twenty-five years later, BGP is still used to direct most of the long-haul communications on the Internet.

As interesting as this "Three Napkin Solution" story is, what does it have to do with cyber security? When Rekhter and Lougheed penned their solution, they did not include security in the design. BGP relies on networks sharing information about what computers they can communicate with. This information is then used to figure out a way to get a message from one computer to another.

I think about BGP as being similar to how I use LinkedIn. When I need to contact a person, I may not know her personally, but chances are, someone in my LinkedIn network does know her and can help me contact this person. Similarly, BGP allows computer networks to tell other computers which computers that network can put them in touch with. This solution is elegant and highly scalable, which is the reason it is still around today.

The problem is that when a network shares information about what computers it can communicate with, there is no way to verify that information. In other words, if a network decides to (i.e., someone makes it) lie about what computers it can communicate with, that network can effectively have data sent to it that it shouldn't see. Similarly if someone in my LinkedIn network indicated that she was connected with Warren Buffet, I may ask that friend to pass a message from me to Warren. This is fine if my LinkedIn contact actually knows Warren Buffet. However, my LinkedIn contact is lying about knowing Warren Buffet, then they have effectively inserted themselves in the middle of a communication from me to Warren. The same thing happens when networks advertise that they can communicate with computers that they can't.

BGP is what directs data over the networks that make up the Internet. If someone can control how that data is directed, he or she can highjack the data as it goes from one computer to another. He or she can essentially be trusted to handle messages that he or she has no business handling and because TCP/IP does not include encryption; the data in those messages can be easily read by anyone that handles them.

This data highjacking is not just a theory. It happens on a regular basis. Pentagon data was sent through Beijing.[8] Data from US corporations was sent through Belarus and Iceland.[9] A Pakistan-based Internet company accidentally took down YouTube.[10] The list goes on, but at the end of the day, BGP is just one technology used to make the Internet work. Numerous other technologies were developed to make the Internet work. Of these technologies, security was not the highest priority, but making the Internet work was the priority. However, I am not blaming the people who helped architect the Internet. To some extent, the Internet is a victim of its own success. If it did not work as well as it does and did not scale as easily as it does, we would not be using it today.

Right or wrong, throughout its history, the Internet was not designed with security in mind. What most people don't realize is that the Internet was not something that a single group of people designed and made in one big meeting. As you can see from this brief history, the Internet is a combination of point solutions that were made by numerous teams and individuals working autonomously. It was not a coordinated effort; it was more organic. As problems arose, different people would step forward to create solutions that would move the Internet forward. This is part of the beauty of the Internet: that something this powerful could be created in such an organic manner is amazing. However, the power and organic capabilities contribute to the complexity of the Internet as well. With no central design or roadmap to coordinate all the pieces that make the Internet work, it is a highly resilient and a very complex entity.

As pointed out in a research paper published by The MITRE Corporation, "One might have thought that computers, their software, and networks were therefore completely understandable. The truth is that the cyber-universe is complex well beyond anyone's understanding and exhibits behavior that no one predicted, and sometimes can't even be explained well."[11] The Internet is a bit like Frankenstein. Different pieces have been put together to create something and at some point, that something has taken on a life of its own. As it has taken on this life, it has done so at the expense of security. As Craig Timberg of *The Washington Post* puts it, "From its unlikely roots in a Pentagon research agency, the Internet developed into a global communications network with no checkpoints, no tariffs, no police, no army, no regulators, and no passports or any other reliable way to check a fellow user's identity. Governments would eventually insinuate themselves into cyberspace—to enforce their laws, impose security measures and attack one another—but belatedly and incompletely."[12] As we can see from this brief look at how the Internet was created, it was not designed with security in mind. What started as a nice way for a handful of researchers to share information has now turned into something that gives all of us the unprecedented ability to communicate around the world. It has grown into a powerful tool, and there are now those who seek to wield that tool for malicious purposes. The question still remains though, why is cyber crime such a problem now? Hasn't the Internet been around for a while? Why wasn't cyber crime an issue from the beginning? What changed that we suddenly hear so much about it?

NOTES

1. "Social Media and the Arab Spring," accessed December 1, 2015, https://en.wikipedia.org/wiki/Social_media_and_the_Arab_Spring.

2. "Estimates of the Impact of Crime Risk on Property Values from Megan's Laws," last modified June 2008, https://www0.gsb.columbia.edu/faculty/jrockoff/aer.98.3.pdf.

3. "ARPANET," accessed December 19, 2015, https://en.wikipedia.org/wiki/ARPANET.

4. Markoff, J. (1999, December 20). "An Internet Pioneer Ponders the Next Revolution." *New York Times*. Archived from the original on September 22, 2008.

5. "Internet Protocol Suite," accessed December 6, 2015, https://en.wikipedia.org/wiki/Internet_protocol_suite.

6. "Net of Insecurity," accessed December 6, 2015, http://www.washingtonpost.com/sf/business/2015/05/30/net-of-insecurity-part-1/.

7. "Net of Insecurity," accessed December 6, 2015, http://www.washingtonpost.com/sf/business/2015/05/30/net-of-insecurity-part-1/.

8. "Net of Insecurity," accessed December 6, 2015, http://www.washingtonpost.com/sf/business/2015/05/31/net-of-insecurity-part-2/.

9. "Someone's Been Siphoning Data Through a Huge Security Hole in the Internet," Accessed Decmber 7, 2015, http://www.wired.com/2013/12/bgp-hijacking-belarus-iceland/.

10. "Pakistan's Accidental Youtube Re-Routing Exposes Trust Flaw in Net," accessed December 7, 2015, http://www.wired.com/2008/02/pakistans-accid/.

11. McMorrow, D. "Science of Cyber-Security." MITRE Corporation, November 2010, http://fas.org/irp/agency/dod/jason/cyber.pdf.

12. "Net of Insecurity," accessed December 6, 2015, http://www.washingtonpost.com/sf/business/2015/05/30/net-of-insecurity-part-1/.

Chapter 4

Why Is Cyber
So Big?—The Technology Curve

The Internet gives power to both individuals and criminals. We are used to hearing about the power of the Internet and how it can improve our lives. It wasn't until recently that we started hearing about cyber crime and how the Internet is being used for malicious purposes. Why did it take so long for criminals to exploit the power of the Internet?

In the mid-1990s companies like Google, Amazon, and Facebook were founded. These companies are all household names because they were able to latch onto the power of the Internet. They were able to leverage the connectivity that the Internet provided in order to improve our lives and bring us services and experiences that we wanted.

Visionaries at these companies pioneered new ways to take advantage of the Internet. Whether it was Google providing us instant access to data, Amazon providing a marketplace that connects buyers and sellers from around the world, or Facebook allowing us to stay connected with our friends and family, each of these businesses found new and exciting ways to utilize the Internet for our benefit.

Let's take Facebook, for example. Founded in February of 2004,[1] by the end of the year, it already had 1 million users. By the fall of 2012, less than 10 years after their founding, Facebook had over 1 billion users.[2] A million users is a lot, but 1 billion is enormous. That is a 1,000-fold increase in users in less than 10 years.

Google, Amazon, eBay, and many other businesses that we consider cornerstones of the Internet all have similar stories of how they grew and changed the way we live. These new businesses and their growth are directly tied to how these companies leverage and used the connectivity that the Internet provided. As they did this, enormous wealth was created for the founders, leaders, investors, and employees of these companies.

Silicon Valley has something called "The Facebook Effect" that describes what happened to real estate prices when Facebook decided to move its headquarters from Palo Alto to Menlo Park. Selecting a relatively inexpensive neighborhood to build their new campus, Facebook essentially drove real estate prices through the roof. In a two-year period, real estate prices in the area of the new Facebook headquarters jumped 41.9 percent.[3] Facebook had generated so much wealth for itself and its employees that they were able to essentially price current residents out of their own neighborhood.

In a capitalistic economy where the pioneers of these Internet companies took tremendous risks, they are now reaping tremendous rewards. Unfortunately, this same risk and reward equation doesn't only work for people who are attempting to improve our lives. It also works for criminals as well. While these companies were leveraging the Internet to change the way we live our lives, there were also people finding ways to use the Internet for more nefarious purposes. The question remains: The big Internet companies were successful and changed our lives relatively quickly, so why are we just now seeing the criminals take advantage of the Internet?

One reason for this is a simple case of funding. Large Internet companies all had access to funding. Google, for example, raised $25.1 million dollars in venture funding before they went public.[4] Fortunately for all of us, the average cyber criminal does not have access to that kind of startup capital. Venture capital funds are not set up to fund criminal activity. This means that the roots of cyber crime had to take hold somewhere else.

But, where else could people find funding to develop ways to exploit the connectivity of the Internet in order to steal information? I'll give you one guess. In 2013, the US government spent over $71.8 billion dollars on intelligence.[5] The United States was not the only country to spend large sums of money on intelligence. If anyone had funds to invest in leveraging the connectivity of the Internet in order to steal information, government intelligence agencies around the world did. As we will see, many of the technologies and techniques used by cyber thugs today come from what governments developed in order to spy on each other.

When discussing any technology and how it changes over time, you need to keep two principles in mind. The first principle is that over time technology becomes less expensive. The second is that with technology, what is hard to do eventually becomes easy. Both of these principles apply to the technologies that we use on a day-to-day basis as well as the ones that criminals use. Let's take a closer look at each of these principles and how they apply to cyber security.

First is the principle that technology continues to become less expensive over time. A classic example of this is something called Moore's Law. In 1965 Gordon Moore, the co-founder of Intel, noticed that the number of transistors (the atomic component of computer chips) that could fit into a square inch doubled every year. Gordon predicted that this would continue into the future. In 1975 this prediction was modified to say that the number of transistors would double every two years instead of every one year,[6] and from that point forward the prediction has been called Moore's Law and has held true to this day. In 1971, the 4004 series microchip had around 2,300 transistors on it. In 2015, Oracle announced the SPARC M7 chip that contains 10 billion transistors.[7] After all this time, Moore's Law is alive and well.

Wait a minute, you might be saying to yourself. What do the number of transistors have to do with technology becoming less expensive over time, and why are we looking at Moore's Law to demonstrate this concept? As more transistors are crammed into computers, those computers become faster and more powerful while the price remains fairly constant. This means that the amount of computing power in an expensive computer today will be the same amount of computing power in a cheap one tomorrow. Over time technology becomes less expensive.

For example, the Commodore 64 that I had growing up in the 1980s was capable of doing .43 million operations per second. In other words, it could not even process half a million instructions in a single second. When it was released, you could purchase a Commodore 64 and a floppy drive to go with it for around $900.[8] Compare this to a 2013 PC based on an Intel Core i7 processor that can do 188,502 million instructions per second.[9] Yes, we went from less than half a million instructions to almost 200 thousand million instructions, and the price of all this additional capability remains relatively the same. When I say that technology tends to become less expensive over time, this is what I am talking about.

Most likely, we have all experienced this trend ourselves. While in college, I sold computers at Best Buy. Just to set the record straight, I was not, I repeat not, on the Geek Squad, not that there is anything wrong with that. I was one of the sales people whose job was to help customers, most of whom had minimal computer experience, navigate the large number of options they had when purchasing a computer. It was a great college job that allowed me the flexibility to only work during the summer and Christmas breaks. Because I came and went throughout the year, each time I came back from being at college for a couple months, I had to re-learn all the computers on the floor. Each time I did this, I was struck by how much more computing capability could be purchased for the same price. The difference was very stark. At one point, I even had a prospective customer comment to me that she had not purchased a computer because of this trend. She said she wanted to keep watching the market and wait until the technology change slowed down. If she stuck to her guns, I'm sure she is still waiting. Technology continually becomes less expensive over time.

The second principle we need to keep in mind when discussing technology is that over time what starts out as difficult eventually becomes easy. When contemplating this principle, I always think of videoconferencing. I first experienced videoconferencing over fifteen years ago. I can remember my first videoconference was for a class project. A team and I were working with a couple of NASA engineers to complete the project. I was a college student at that time, and the only thing cooler than having a videoconference was having a videoconference with people at NASA. The conference was the culmination of a project that my classmates and I had been working on for months. There was a lot of anticipation on our part.

When the day came, we all met in a special room that had been outfitted with a videoconferencing unit. Not long before this day, the college had spent lots of money creating this room. To be clear, this room wasn't just another computer lab where students could come, try the technology, learn how it worked, and then go about their way. No, this room was in a corner at the end of a hallway and the door was always locked. Apparently, the college was very proud of this room, and it was only to be used in special circumstances and under very strict supervision.

For this special meeting, we all arrived early just to make sure we had time to handle any last-minute issues. It's a good thing we arrived early because it took us about fifteen minutes to get the conference set up and working. Finally, the images of a NASA conference room began to appear on our TV screen. It was amazing. As the NASA engineers entered the room, we exchanged quick introductions and jumped right into our project. At this point the video started to freeze. The images of the engineers froze with their faces in the most unnatural positions: mouths wide open, eyes half shut, looking as though they were about to sneeze. As we started wondering what we might be looking like on their end, the video suddenly unfroze and jumped back to life. Looking back on this experience, as fun as it was "holding a videoconference" with NASA, the meeting really wasn't as productive as I anticipated.

Compare that experience to another one I recently had. While visiting San Francisco for work, I found myself with some free time and decided to explore Highway 1 along the coast. A colleague and I happened upon a high cliff overlooking a beach with surfers riding the waves in. To me, it was the perfect picture of California, and I wanted to share it with my wife and kids. Rather than taking a picture, I decided to call my wife's cell phone. When she answered the call, my smartphone popped up an icon with a video camera on it. I touched that icon and oh look! I instantly started a videoconference. I was not in a special room. I did not have any special training—an instruction manual may have existed explaining how that feature worked, but I'm a guy and I can assure you I didn't read the manual. There I was standing on a California beach letting everyone at home watch the surfers along with me.

I should also point out that my wife took advantage of my absence, and she and the kids were at an amusement park when I called. I was able to see the kids playing and having fun at the park. My point is that neither of us was planning on holding this impromptu videoconference, and the only thing we needed to hold the conference was a piece of technology we both carried in our pockets. What was once hard has now become easy.

You probably will not share my videoconferencing experience, but I am fairly certain that you have experienced technology becoming easier over time. Do you remember the days of the dial-up modem and America Online? For a while, this was how my family "accessed" the Internet. You would have to set up your modem, pick the phone number you would try to connect to, hope that the line would not be busy, hang up and dial the number again because chances were the line *was* busy, and after several attempts, finally you could connect. But wait, you were not yet on the Internet. After connecting, you would have to find the web browser and load it. This would finally allow you to "surf the web."

For those of you who never experienced this, yes, the Internet was something that you actually had to access. There were no "apps" that stored your data in the "cloud." Your device did not automatically "connect" when it turned on. In fact, most of the time, my computer was not connected to the Internet. It was only when I specifically wanted to do something on the Internet that I connected.

Today, the Internet is ubiquitous. It is everywhere and all of our devices are on it. Even your child's Barbie doll, your refrigerator, and the thermostat in your house are now on the Internet. It is so easy to "connect" to the Internet that most of the time, we do not even think about it. The Internet is just there. This was not always the case. As time goes on, technology becomes easier.

As we can see, all technology follows these two principles. Every technology will become less expensive and easier to use over time. It's not just accessing the Internet, or videoconferencing, or ever increasing computer horsepower that follows these principles. If you think about it, all technology follows these principles. What is more, helpful and useful technologies aren't the only ones that follow this curve. Malicious technologies—technologies used to execute cyber attacks—also follow this trend.

When I worked in the defense industry, we spent a lot of time stopping governments from breaking in and stealing secrets. The attackers we faced were well funded and used the latest technologies in their attacks. These technologies were very targeted. They were developed or customized to work against my company.

Recall from earlier chapters that we discussed how attackers can write custom malware to attack specific victims. Previously the same malware was used to attack as many computers as possible. Our antivirus programs do a good job protecting us from these attacks. By writing programs that are used to attack only a few victims, attackers will usually be able to evade antivirus. From an attacker's perspective, the downside of this approach is that it is expensive to write or modify programs for each victim. This is why, at first, only governments or other well-funded hacker groups would adopt this approach.

Well-funded government operations would be conducted in the interest of national security. In this situation, the price is not the first consideration, so expensive approaches can be taken to ensure success. If this means that custom software needed to be developed to infiltrate a cyber target, then a successful breach was worth that cost.

Over time the two principles we discussed began to apply to hacking technology. What was once hard and expensive eventually became cheaper and easier. At some point, technology became cheap and easy enough for organized crime to use it for financial gain. Once the economics of these types of attacks reached that point, we started seeing attacks, such as the retail store breaches, where the goal isn't to protect national security or steal state secrets, but to steal something that can be sold on the black market for profit. The technology used to execute these attacks became cheap and easy enough that criminals could use it to make money. In other words, as this hacking technology followed the two principles we have been talking about, the cost and simplicity of pulling off highly targeted attacks that included custom software became cheap enough that attackers could still make a healthy profit from selling the data they stole.

Once this happened, cyber attacks moved from the realm of something that only countries do to something that sophisticated criminals do. This is when retailers and health providers suddenly found themselves in the crosshairs. Credit card numbers and medical records have a value on the black market, and those companies suddenly found themselves having to defend against the kind of attacks that previously only government contractors and large financial institutions had to defend against.

Companies didn't see this coming and they didn't plan for it. Because the companies that earlier had defended themselves against nation states had not spoken out about what they were seeing, these other industries were completely surprised and unprepared. Had we thought about how all technology becomes cheaper and easier to use over time, we could have foreseen that advanced hacking techniques would eventually make stealing personal information on an industrial scale a lucrative endeavor that many people would not be able to pass up.

Throughout my time in the cyber security space, I have seen these trends play themselves out many times. In chapter 5, we will take a look at a couple details of an attack to better see how these two principles play out in the cyber security realm. We will look at how these new advanced attacks take place. We will understand how the attackers pull off the large data breaches we hear about in the news. We will also see how the two principles of technology becoming less expensive and easier to use over time apply themselves.

NOTES

1. "Facebook," accessed March 18, 2016, https://en.wikipedia.org/wiki/Facebook.

2. "Number of Active Users at Facebook over the Years," accessed March 18, 2016, http://news.yahoo.com/number-active-users-facebook-over-230449748.html.

3. "The Facebook Effect (On Real Estate Prices)," accessed March 18, 2016, http://techcrunch.com/2015/04/28/the-facebook-effect-on-real-estate-prices/.

4. "Venture Capital Funding: How much VC money did Google raise before going public?" accessed March 18, 2016, https://www.quora.com/Venture-Capital-Funding/How-much-VC-money-did-Google-raise-before-going-public.

5. "You Ask, We Answer: How Much Do We Spend On Intelligence?" accessed March 18, 2016, https://www.nationalpriorities.org/blog/2012/03/16/you-ask-we-answer-how-much-do-we-spend-intelligence/.

6. "Moore's Law," accessed November 29, 2015, https://en.wikipedia.org/wiki/Moore%27s_law.

7. https://en.wikipedia.org/wiki/Transistor_count.

8. "Commodore 1541," accessed November 29, 2015, https://en.wikipedia.org/wiki/Commodore_1541.

9. "Instructions Per Second," accessed November 29, 2015, http://gaming.wikia.com/wiki/Instructions_per_second.

Chapter 5

Anatomy of an Attack

"Sophisticated attacks" seem to occur everywhere these days. Just how "sophisticated" are these attacks? If so many companies are becoming the victims of these attacks, can they really be that sophisticated, or is this just a convenient excuse for businesses that lose data? To answer these questions, we need to take a closer look at how such attacks actually happen. As we do that, it is important to know that although there are similarities between attacks, no two attacks are exactly the same. With that said, there are several steps that attackers must take before they can successfully compromise their target. How they accomplish these steps vary, but all attacks have to go through each of these steps in some way, shape, or form. By looking at these steps, we can better understand just how sophisticated these attacks are and gain a better understanding of what defenders are up against.

After wrestling with well-resourced, state-sponsored attackers, members of the Lockheed Martin Computer Incident Response Team (LM-CIRT) started to see patterns in how attackers went about their attacks. The members of this team pulled these patterns together to create a model that describes the major steps attackers go through in order to compromise a company or individual. This model was published in a white paper and given the name Cyber Kill Chain®.[1] In the cyber security industry, the model is affectionately known simply as the "kill chain." Think of it as a high-level process that attackers follow. We will use this model to help us understand how attacks are executed and what attackers do to get past their victim's defenses.

At a high level, the steps that an attacker goes through to execute a successful cyber attack are as follows. First, the attacker performs "reconnaissance," "recon" for short, on their target trying to find a weakness in the victim's defenses that can be used to gain entry into the computer environment. Second, the attacker uses that recon to create some kind of "weapon" that can break through the victim's defenses. This weapon is usually a custom-made malicious program, just like the custom malware we talked about earlier. With their weapon in hand, the attackers will now "deliver" that weapon to the victim. Like a cyber missile flying through the Internet, the attackers deliver their malicious payload. Once the weapon enters the victim's computer environment, it now has to successfully "exploit" the weakness found in the recon phase—the same weakness that the weapon was custom made to exploit. If the weapon successfully exploits the weakness in the victim's defenses, then the attackers effectively have a computer program that they made running inside the victim's environment. In other words, with a successful exploit, the attacker has been able to successfully break through the victim's defenses.

With the defenses breached and a malicious program that they made now running inside the victim's environment, the attackers will use that malicious program to "install" itself on the computer it is running on. I'm sure we have all experienced the frustration of working on a document only to have your computer crash and realize you didn't save it. All your hard work disappears. The attackers have the same issue. Their weapon can get a program running on a computer, but if they do not install that program so that it will start running again if the computer is turned off and back on again, then they run the risk of losing all their hard work if the computer crashes or reboots. Installation is the attackers' version of saving your work. With their malicious program successfully running and set up to keep running, the attackers now have a stable platform inside the victim's environment from where they can operate.

With this platform in place, the attackers will then establish "command and control." Up to this point, the malicious program that the attackers created has had to do everything on its own. Think of it like a drone that is preprogrammed to exploit and install itself automatically. Command and control is when the malicious program reaches out and establishes communications with the attackers themselves. Once this communication channel is set up, the automated pieces of the malicious program shut down, and the attackers can now directly control everything the program does. In other words, the attackers are now controlling a computer inside the victim's network.

Now that the attackers have control of a computer, they are free to take "actions on objective." They can now seek out the information they are looking to steal, they can plant additional backdoors in the environment, they can steal user IDs and passwords, they can deleted or change data, and the list goes on and on. In short the attackers can take whatever actions they need in order to achieve their objectives.

Recon, weaponization, deliver, exploit, install, command and control, and actions on objective—these are the high-level steps an attacker takes in order to successfully execute a cyber attack. Let's take a look at each of these in more detail and give some examples of how they play out in the real world.

To begin with, the model points out something that most people do not realize. Sophisticated cyber attacks are not random; they are targeted. In other words, the target of the attack is selected, and the attack is customized to evade that target's specific defenses. This means that the first step in the process is to perform recon. This recon helps attackers understand whom they are going to attack and how they are going to attack them. For example, if one government wanted to gather information on a new weapons system that another government is developing, attackers would have to first do some recon to find out who has the data they are looking for. Do any research institutions have it? Do any private companies have it? Maybe a particular government agency has the information they are looking for.

Once they select a target, the attackers then decide how they will go after that target. Would an email-based attack be best? Would it be easier just to attack its web servers? There are a number of ways to execute an attack. Which approach looks most promising and, quite frankly, easiest?

Because these attacks are highly targeted, a fair amount of recon often goes into planning the attack. Understanding which organization to attack is key, but an attacker also needs to understand who works in that organization. Where do they live? What type of work do they do? Who are they connected with? All this information may be needed as the attack progresses.

Once this recon is done, weapons, or custom tools, need to be created. These tools are the malicious programs that will be used during the attack. As the attack progresses, different tools will be needed. In fact, an entire arsenal of tools will be needed. If you have ever seen the 2001 movie *Ocean's Eleven*, starring George Clooney and a full cast of other stars, you can appreciate all the tools needed to pull off a sophisticated heist. In the movie, George Clooney's character, Danny Ocean, assembles a team of thieves to breach the vault of the Bellagio Casino. This vault has many layers of security and is considered impenetrable. To get into this vault, Danny Ocean and his team use a full set of tools, including explosives, a device that generates an electromagnetic pulse, and a remote-controlled van, just to name a few. As the movie unfolds, it seems that Danny and his team always bring just the right tool at just the right time. Although less glitzy, sophisticated cyber attackers operate in much the same way. They have a full set of tools so that as the attack progresses, they have just what they need when they need it. After attackers have finished their recon, they then prepare all of these tools as part of the weaponization phase.

Danny Ocean and his crew used remote-controlled cars and explosives. What kind of tools do cyber attackers use? Many times attackers will use an email-based attack to get some unsuspecting person inside their target to unwittingly install what is called a "backdoor" on his or her computer. The backdoor is one of the key tools that cyber thieves use. It essentially allows an attacker to take over a computer. Numerous types of backdoors exist, but a common example is a program that attempts to make a connection *from* the victim's computer *out* to the attacker. The key here is that the backdoor, once installed on a computer inside the target, will initiate communication out to the attacker.

This strategy is brilliant. For the most part, companies do a really good job stopping communication from coming *into* their computers. It is pretty hard to come from the Internet and initiate a conversation with a random computer inside a company. However, it is relatively easy if you are on a computer inside a company to initiate a conversation with a computer on the Internet. Based on this, if an attacker can get his or her backdoor installed on a computer inside a company, the backdoor can open the proverbial door of the network from the inside, thus the name backdoor—as in a backdoor into a network. This gives the attacker a much better chance of gaining access to that network.

This backdoor is just one example of a tool that the attackers would create at this step in the attack process. There are many other tools, such as a tool that will help get the backdoor installed, a tool that finds the information the attackers are looking for, as well as tools for stealing passwords, and many others. Attackers create a full toolbox of weapons that they can draw from as they progress through the attack.

Once the research is done and all the weapons are created, it is time to launch the opening salvo at the target. This often means delivering something designed to exploit a weakness in the victim's defense in order to get the backdoor installed. Remember that the backdoor allows an attacker into the target's network, but the first trick is getting that backdoor installed so it can do its job. There are many ways to exploit a weakness. One of the most popular approaches used by attackers is to send a well-crafted email that essentially exploits people, tricking someone inside the target company into unwittingly installing the backdoor. These emails are not the typical fraudulent emails you may have seen from Nigerian princes with misspelled words and bad grammar. These emails are very well put together and designed to make people open them.

On the high end, I have seen—during the 2010 BP Horizon oil spill in the Gulf of Mexico—people receive an Environmental Protection Agency (EPA) water quality reports for vacation homes that they owned in the area. The water quality report was an Adobe PDF file that had some malicious code in it that installed the backdoor when the file was opened. A couple things struck me about this email. Yes, the attackers created a PDF file that was able to install a backdoor. That is interesting, but honestly happens quite often.

What really stood out was that the attackers knew that this person owned a vacation home in the vicinity of the spill. They did their research well, and all they needed was for their victim to click on the water quality report, and their backdoor would then be installed.

I've also seen engineers attend conferences, and then a week later receive an email from "the keynote speaker" thanking them for attending the conference and attaching a copy of the slides used during the presentation. The engineer thought "the keynote speaker" was just a nice person who was really good at following up. In reality the email did not come from the keynote speaker, but from an attacker, and the slide deck that was attached had malicious code in it just like the EPA water quality report did that installed a backdoor on the engineer's computer. My point is that the really good attacks are not based around Nigerian princes who want to send you money. Some of these attacks are very customized and created to appeal to their audience.

The research, weapon or tool development, and effort to get that tool delivered all needs to come together in order to get that backdoor installed on the victim's computer. If attackers can get that backdoor installed, then a big part of the compromise is complete.

Once the backdoor is installed, it then initiates a conversation with another computer that is on the Internet. This second computer is a machine that the attackers already control. When the victim's computer initiates the conversation, it goes something like this: "Hi. I'm in. What do you want me to do?" In essence, the backdoor establishes a channel through which the attacker can now control the victim's computer. Once this happens, the victim has no idea, but he is in reality sharing his computer with an attacker. Cyber security professionals call this "command and control." The attacker basically has command and control of at least one computer inside the victim's computer network. Once command and control is established, the attackers can bring in tools that allow them to dump user IDs and passwords.

In most companies, there are central servers where all user IDs and passwords are stored. When you enter your user ID and password, these servers check what you entered and tell your computer whether you entered the right information or not. These servers are juicy targets for attackers. If the attackers can get their hands on the user IDs and passwords, they can then log into any computer in the company just like an employee would. Using the user IDs and passwords, attackers do not have to hack. They just need to login like everyone else.

In order to get these user IDs and passwords, attackers have developed a handful of tools. These tools allow them to dump all the passwords on the central servers. Once they have the user IDs and passwords, attackers can then move around the other computers inside the company until they find what they are looking for and eventually take it.

This part of the process was highlighted in the 2014 Sony Pictures Entertainment hack. As information about the attack came out, many people thought an insider, someone who worked at Sony, had pulled off the attack. Several of the tools used in the attack contained the names of specific Sony servers as well as user IDs and passwords to those servers. This means that whoever pulled off the attack somehow knew not only the names of Sony's internal servers, but also the user IDs and passwords that gained access to those servers. This is obviously information that someone working at Sony would know. Many people believed this assumption and asserted that the attack had to be done by someone working at Sony. As more information came to light, it became evident that the attackers were not people who worked at Sony, but that the attackers were just in Sony for so long that they not only got to know all the Sony servers and what was on them, but that they had also stolen the user IDs and passwords needed to gain access to those servers. The Sony hack was not an inside job. The attackers just stole the user IDs and passwords and used them to move around the Sony computers just like an insider or an employee would.

As you can see, there are several steps that need to be taken before an attacker can successfully execute an attack. Each step is relatively complicated and takes special tools and capabilities to pull off. In all transparency, the descriptions above are simplified versions of what really happens. There is much more detail in each of the steps in an attack. I intentionally simplified them so we did not get lost in the minutia. Even at this high level, I'm sure you can appreciate how complicated these attacks are.

In this sense, the attacks are sophisticated. However, as we apply the principles we discussed in chapter 4—that technology becomes less expensive and easier to use over time—we will see that even though there are multiple steps in these attacks, as well as multiple tools needed to pull them off, they are becoming less expensive and easier to execute. When these kinds of attacks were first used, they were very expensive and very rare. These were the types of attacks I was seeing from countries trying to steal state secrets.

Fast-forward ten years. The technology behind these attacks has become less expensive and much easier to use. Tools now exist that can automate parts of this attack process. Software can be purchased on the black market that can make most of the steps in the kill chain easier. Let's look at one tool in particular to see how these attacks have become less expensive and easier to execute.

Many of the sophisticated attacks I have seen use a lot of well-crafted emails to get a backdoor installed. These emails look an awful lot like the EPA water quality report in the fraudulent email I mentioned earlier. Attackers craft some kind of an email that has an interesting attachment. When the victim opens that attachment, such as a Word document or a PDF, a piece of malicious computer code executes and installs the backdoor on the victim's computer. But, how can this be done and not caught by security tools? The key to making this work is referred to as "packing." The trick to packing malware is that the attacker has to embed the backdoor inside another file, oftentimes the attachment sent in an email, like the EPA water quality report, in such a way that it couldn't be detected but that would still allow it to automatically be installed. In other words, the attacker is taking one computer file, hiding it inside a second file while also crafting a trigger that will automatically extract and install the first file once the second file is opened. This process is the cyber equivalent of those Russian nesting dolls that hide a smaller doll inside a larger doll. If this sounds complicated, that is because it is complicated.

A handful of techniques exist for doing this, but none of them are very easy and most of them involve in-depth understanding of how computers use memory along with some fairly complicated math to pull off. The first few people to make this work were not only very smart, but they also had a decent amount of time to spend figuring it out. In other words, it was very expensive and very hard. Insert our two principles; over time technology becomes less expensive and easier. Eventually, the smart people who knew how to successfully pack malware developed small utilities that automated the packing process. Once these utilities where developed, the attackers no longer needed advanced degrees in computer science to embed their backdoors inside other files. They just needed to know how to use the utility.

I can remember one instance when my team caught a malicious email. Very similar to the EPA water quality report or the presentation slides from "the keynote," this email had a PDF file as an attachment. Taking a closer look at the PDF, we found that another file was packed inside of it. When we find packed files, those packed files are almost always backdoors. This time was the exception to the rule though. As we pulled the two files apart, we found that the malicious program that was packed inside the PDF was not malicious at all. The embedded program was the packing utility. It was the tool someone had written in order to make the packing process easier.

When we fired up this utility, we saw exactly how attackers packed their malicious programs inside other files. The utility opened a small window where the attacker selected the file containing the backdoor that was to be packed as well as the benign file that would house the backdoor. Once they selected their two files, the utility would package up the backdoor, perform all the complicated work of embedding it in the benign file, and save it out for the attacker to use. Rather than needing a brainiac to pack backdoors inside another file, the utility made it so easy that anyone could do it. In this case, the attacker was so inexperienced he didn't realize that instead of embedding a backdoor, he had embedded the actual utility itself. As you can see from this example, what was once expensive and hard to do became cheap and easy.

Now that utilities like this one have been created, the cost and complexity associated with developing and executing highly targeted attacks has come way down, so far down that we are seeing nongovernment-related hackers execute these kind of attacks for financial gain. As I have worked across several industries and with numerous companies, I increasingly see small suppliers of larger companies breached in order for the attacker to gain access into the larger company. In these situations, the supplier didn't have anything that the attacker wanted, other than a trusted relationship with the larger company. Hacking into the supplier was just a step along the path of hacking into the larger company. That's right, these attacks have become so easy that attackers do not hesitate to compromise two companies in order to get to the data they want. This situation may sound bad, but it gets worse. The trend that has made hacking technology cheaper and easier to use will only continue. This means that more and more people will be able to use this technology for more and more reasons. The rise in hacktivism is one example. Just like activists try to make a statement and draw attention to their cause, hacktivists are people who use hacking to make their statements. If they don't like something a company has done, rather than protesting outside the front doors of the company headquarters, they take down that company's website. Why stand in the rain and snow when you can make your voice heard from the comfort of your home?

One favorite tool of hacktivists is something called a distributed denial-of-service attack (DDoS). This DDoS is essentially when an attacker sends so many fake requests to a website that the website cannot keep up. If the volume of requests is sustained long enough and at high enough levels, the website will eventually go down. These types of attacks used to be very sophisticated and hard to pull off. Now, they have become so easy that individuals can download an app to their Android phones and participate in an attack. It's so easy; a lot of people don't even realize they are participating in something harmful and illegal. Now that it is so easy to join in on a DDoS attack, hacktivists no longer need to recruit highly talented, technical people to pull off their attacks. They can simply attract anyone who believes in their cause and can download an app.

As this trend continues, cyber security will become more and more of a problem for all of us. Are you mad at your boss? Hack him or her. Did your city council make a decision you don't agree with? Steal all his or her emails and publish them on the Internet. Did your favorite coffee shop raise its prices? Bring down its website.

It is a brave new world, but it is not the end of the world. The history of how the Internet grew up as well as the universal trends of technology becoming cheaper and less complicated have come together to put us in a challenging spot when it comes to protecting our personal and business data.

We can see how we have arrived at this spot. Now let's take a look at how we can move forward. Let's examine how businesses can protect their most sensitive data and how individuals can protect themselves and their private information. We will see that just as technology has changed the way we work, do businesses, and interact with other people, it should also change the way we approach security.

NOTE

1. "Intelligence-Driven Computer Network Defense Informed by Analysis of Adversary Campaigns and Intrusion Kill Chains," accessed December 5, 2015, http://www.lockheedmartin.com/content/dam/lockheed/data/corporate/documents/LM-White-Paper-Intel-Driven-Defense.pdf.

Chapter 6

Businesses—Different Risks for Different Businesses

The current state of cyber security is not good. Every day we feel as though another organization has been hacked and its data stolen. The question remains: What should we do to stop it? As with most complicated problems, there's no one-size-fits-all answer or a sure-fire way to solve the problem, but there are very effective ways in which we can approach the problem.

Cyber security is a risk. Just like the risk of a stock market crash, the risk of a war, the risk of a competitor leapfrogging your technology, or even the risk of an earthquake, cyber security and the data breaches associated with them are risks that businesses must assess and plan for. Fortunately, cyber risk is not the first risk that people and businesses have had to deal with. This means that there are models out there that can help guide the way we look at cyber risk and how we protect our businesses and ourselves from it.

Just like the cyber security world, the financial investment world is a high stakes and high-risk environment. Just as a guaranteed strategy for full cyber protection does not exist, nor does a guaranteed strategy for making money through investments. Despite what the advertisements of many financial institutions may lead you to believe with pictures of retired couples on beaches, profiles of large stately buildings, or even images of large solid rocks, returns are never guaranteed on investments. Each investment carries a certain amount of risk, and the return on that investment should, in theory, be in line with the risk. The theory is that the bigger the risk, the bigger the possible return on the investment.

After many years of intense study, some standard approaches have emerged to help guide investors and increase their chances of success. From my experiences, I believe that cyber security professionals can learn a lot from what the investment community has already done, particularly when it comes to profiling risk.

When dealing with investments, different people have different risk profiles. If risk is a continuum from high risk to low risk, a retired investor will generally take less risk than a young investor just entering the market. Why is this? Why are younger investors generally willing to take more risks than older investors? It is not because younger people don't care or tend to make hazardous decisions, although that may be true to some extent. No, younger investors tend to take more risk with their investments because they have a longer time horizon on when they expect to actually need their money. For the vast majority of people, the reason they invest is so they can enjoy retirement. This means that younger people are not expecting to sell their investments for twenty or thirty years. With this kind of time horizon, they can afford to shoot for the higher risk, higher return investments. For them, the prospect of losing money is not that bad. If they lose, they still have time to make it up.

Compare this to older investors who are very close to retirement or even in retirement. For them, the prospect of losing money is a much bigger concern. If you are retired and lose all of your retirement savings, you have to go back to work. That kind of defeats the purpose of investing, doesn't it? It is this risk versus reward calculation that drives investment decisions. Realizing this, many investment advisors have adopted several risk categories to help them understand how each of their clients evaluates risk versus reward.

One cornerstone of modern investment strategy involves creating investor categories. These categories help investment advisors understand the risk profile of their clients. Understanding this risk profile gives financial advisors a framework for making investment recommendations. For example, financial advisors may recommend low-risk bonds to conservative investors, while they may also recommend growth stocks to aggressive investors. Bonds do not generally make sense for aggressive investors because returns are not very high, and growth stocks do not generally make sense for conservative investors because they are too volatile.

Several years ago I switched investment advisors. I was early on in my career, didn't have much in my account, and as a result didn't get much attention from my original advisor. As I searched for a new advisor, I spoke with a number of very credible and highly recommended people. After a couple of conversations, I spoke with an advisor who didn't just talk about what I would need to do to put my children through college and retire comfortably. No, this person wanted to know primarily about how I viewed risk.

Don't get me wrong, financial goals, such as sending children to college and retiring comfortably, were part of the conversation. However, most of the conversation centered around how I viewed risk. Would I take a big risk in order to get a big immediate gain? Would I take a big risk in order to get a big gain twenty years from now? What if I could make an investment that lost money in the short term but had the potential for a big upside in the long term? In what situations would I be more inclined to be aggressive? In what situations would I be more risk averse? What situations would cause me to take on more risk? We talked about a number of scenarios. The goal of this exercise was to determine how I viewed risk and how that should play into my investment strategy. Based on my view of risk, this advisor put together an investment strategy that fit my goals. In other words, my goals did not drive the strategy; it was my view of risk and the type of investments that would lead us toward those goals that drove the strategy. My view of risk was the foundation for future investment decision making.

Having gone through this process, it struck me that this concept holds true in cyber security. One company may be more risk averse than another company. A certain business opportunity may cause a business to take more risk than it normally does. Other situations may cause a business to be more risk averse than it normally is. Just like investors and their view of risk, companies view risk and evaluate risk differently.

The cyber security professional's job, just like the financial advisor's job, should be to understand their businesses' view of risk. Cyber security leaders need to know how their businesses understand, view, and evaluate risk. They need to know what business opportunities will cause the business to take more risk and what situations will cause it to take less risk. Cyber security leaders need to understand that risk is not a zero-sum game. Risk is not black and white. We do not live in a world where risk either exists or does not exist, and the cyber security leader's job is not to obliterate risk where it does exist. No, the job of the security leader is to understand the broader risk and reward equation for the business. They need to help the business fully understand the risk side of that equation. They need to educate other leaders as to the risks that exist in the cyber realm, but they also need to understand that if the reward is high enough, the risk needs to be taken. As leaders within businesses, cyber security professionals need to understand this.

As I mentioned previously, I did not always want to be a security leader. One of the reasons I did not aspire to the IT security profession was a perception I had of security leaders. As many parts of the business were trying to make things better and move the business forward, IT security seemed to be a constant roadblock. They had a very black-and-white view of the world. If a project introduced any risks, those risks must be removed or else the project simply could not go forward, regardless of how much value that project would bring to the business.

Before I was a security leader, I ran a server, storage, and network design organization. I was basically in charge of the design of all the large pieces of hardware that ran all the business applications. The business spent well over $100 million dollars annually keeping all these pieces of hardware up, running, and current. The cost to the business was nothing to sneeze at. In order to free some of these funds up so they could be used to help grow the business, my team and I were hoping to introduce a new technology called "virtualization." With this technology, we could better use the investment we had already made and save a considerable amount of money going forward. With this technology, we planned on delivering tens of millions of dollars back to the business.

As the team went about implementing this new technology, we came to the obligatory "security review." This is when the IT security team reviews the project to make sure new security problems are not being created. This was supposed to be a check and balance in the process to ensure people were not "riding fast and loose" when it came to security. Unfortunately, the pendulum swung too far the other way, and these meetings often turned into a bit of a gauntlet that had to be run. The security team held a veto on any project, and if they perceived that a project was introducing risk into the environment they did not hesitate to use that veto to block a project.

Now don't get me wrong, as a former cyber security leader, I am not saying security teams should not have this veto power. They should. What I am saying is that the security teams need to be judicious in how they use that power. They need to understand the broader risk versus reward picture for the business before using their veto. They also need to explore all options with the business to reduce risk before using their veto.

In the case of my virtualization project, I can distinctly remember the conversation during the security review. At the time, virtualization was a relatively new technology. It held a lot of promise and was gaining a lot of support, but it was not yet considered an industry standard as it is today. As we met with the IT security team for the review, one of the members of the security team began painting a picture of how this technology could be exploited. I do not recall the exact logic, but it was something along the lines of "if someone could do A and if they could do B and if they could do C at the same time and then after that do D and E, then it is possible, in theory, that they could break in and steal information." Based on this "if someone could" logic, the security team began moving to veto my project. A perfectly good project that would return tens of millions of dollars back to the business was about to be shut down because of a theoretical risk.

Honestly, the reason I don't recall the logic behind the perceived risk the security team was telling me about is because after three or four "if someone could" I stopped listening. When the security team member finished his dissertation, I simply asked whether the hypothetical scenario he just outlined had ever happened, whether in a lab or in the real world. As a point of fact it had not. With this fact established, I then asked whether they were prepared to waste tens of millions of dollars on a hypothetical risk that had never manifested itself. The answer was yes. They felt like my project was introducing a new risk into the environment and from their perspective, any new risks were unacceptable. In my opinion, there was a complete lack of understanding risk versus reward. The business had to take calculated risks to stay competitive.

The security team only looked at the risk side of the equation. The project contained risk, and regardless of how big or small that risk was, the risk existed and had to be eradicated. In this case, eradication meant the killing of my project. For me, this experience simply put a bad taste in my mouth, and I believe there are a number of people out there who have had similar experiences with other IT security teams. As a side note, the virtualization project did eventually move forward and was implemented. The virtualization technology used in it is now a standard in the industry and used almost everywhere.

Ultimately, security teams need to be like my financial advisor. They need to understand how the business views risk versus reward. They need to educate the business on the risks associated with cyber, but they also need to appreciate that the business must take risk in order to compete. Risk is not black and white, but it is gray. Whereas IT security leaders tend to live in black and white, successful cyber security leaders have to live in the gray. Successful cyber security leaders need to understand how their businesses deal with the gray and where they draw the line. Successful cyber security leaders need to be able to educate the business on cyber risk while also doing all they can to reduce that risk and stack the odds of a good return in favor of the business.

To understand how businesses view risk versus reward, it is helpful to create a couple of cyber security risk categories that we can fit businesses into. These categories will give us a reference in which we can place different businesses.

For the sake of brevity, we will use two broad categories. The first category consists of businesses that have commodity data, data that many other businesses have as well. Whether this data contains customer credit card information or personal information, such as medical records, a number of businesses out there have the same kind of data. Taken as a whole, what these businesses store is not entirely unique. The individual credit card numbers or healthcare records may be tied to a unique person, but the fact that the businesses have credit card numbers and healthcare records is not unique.

The second category that we will explore is businesses that have unique data. Most of the time, this takes the form of sensitive intellectual property. These businesses may invest heavily in research and development. They might be trying to develop the next best widget. The future of their business often depends on them being able to sell enough of those widgets to not just recoup their investment in research and development, but to also turn a healthy profit. For these businesses, losing the data that only they have to a competitor could result in the death of their business.

Before we get too far ahead of ourselves, let's look more closely at our first category: businesses that have commodity data. Home Depot, Living Social, and eBay are just a few such companies that have experienced a breach. Whether they realize it or not, customers trust these businesses to protect specific pieces of their information whenever they shop at their stores or use their services. Credit card numbers, Social Security numbers, cell phone numbers, mother's maiden names, passwords—the list of customer-sensitive information goes on. How should these businesses look at and respond to cyber risk?

To answer this question, you must understand why these types of businesses are the targets of cyber attacks in the first place. Are hackers stealing credit card numbers because they want to go on a spending spree? Are they stealing your mother's maiden name because they think it is interesting? Not exactly. In most cases, hackers steal customer information to sell it on a black market. For example, credit cards are stolen and sold online in what is known as "carding forums."

Most people are not aware of the less savory side of the Internet where these kinds of websites exist. You cannot search this side of the Internet from Google, and in most cases you need specific software on your computer before you can get to it. This special software attempts to provide anonymity by hiding who you are, what computer you are using, and where you are coming from. With your identity sufficiently hidden, you can access the dark underbelly of the Internet where you can purchase everything from stolen credit cards to illegal drugs, and even contract a hit man.

Popular destinations in this part of the Internet are sites called "carding forums." These sites are places where carders—people who actively trade stolen credit card information—gather to buy, sell, and trade their wares. As soon as you access a carding forum, you can purchase stolen credit cards. Buying a stolen credit card is similar to buying a book on Amazon. These sites have a shopping cart that you can fill with your purchases and then proceed to checkout. The only difference is that you don't pay with a credit card, but with Bitcoins. Bitcoins are an electronic currency that provides some layer of anonymity.

These sites provide a ready market upon which cyber criminals can sell stolen customer information for profit. It is in these black markets that most of the customer information stolen from companies is sold and the attacker gets their payday. Similarly, cyber thieves are beginning to steal not just credit cards, but information that can be used for identity theft. Credit cards eventually expire and can be turned off, but stealing a person's identity is the gift that keeps on giving. Purchasers of identities on these black markets can create new accounts and file false tax returns years after they steal the original information.

You can see the repercussions of identity theft with the increase in tax fraud. The IRS reported a 66 percent increase in fraud investigations from

2012 to 2013.[1] The Government Accountability Office estimated that fraud tied to identity theft cost taxpayers $5.2 billion.[2] The problem got so large that in 2015, TurboTax had to temporarily stop filing state returns for their customers.[3] Most people find that they are the victims of identity theft when the government rejects their tax filing because someone already filed a fake tax return before them.

Most people don't realize just how deep this illegal activity goes. An entire economy has developed around selling stolen information. For example, purchasing information that can be used for identity theft is much more expensive than purchasing credit card numbers because an identity has more long-term value associated with it.

Understanding this concept greatly impacts how companies that protect this information should approach cyber security. Understanding that data is a commodity that can be bought and sold gives us a clue as to how we should approach the problem. As with any commodity, such as oil or silver, the people bringing these commodities to market are going to try to obtain them as cheaply and easily as possible.

The natural gas and oil industries illustrate this point. Both natural gas and oil are commodities that are bought and sold in an efficient market. There are many ways to obtain them. Looking at some of these ways, we can see that offshore drilling is obviously harder and more expensive than land-based drilling. The first developed natural gas well in the United States was discovered in 1815.[4] It wasn't until 1947 that the first offshore well was drilled out of sight of the shore.[5] Why did it take over 100 years for this advancement? The reason is because there was a large supply of easily accessible oil and gas on shore. It wasn't until the economics changes that oil and gas companies found it economically feasible to drill offshore. Why would oil and gas industry leaders spend more time and money to obtain their product than they had to?

Stolen commodity data works the same way. Remember, hackers pilfer commodity data so they can sell it. That's right, data is just another commodity that is bought and sold to the highest bidder. Even before the advent of cyber attacks (that lift thousands of credit card accounts or other personal information) were conceived, carders still obtained this information and sold it for profit. Whether it was the waiter who wrote down your card number while you were paying at a restaurant or just a thug who held you up and took your card, people have been stealing credit card numbers for a long time now. But the game has changed. Over the last couple of years, the tools and technology to steal commodity data via cyber attacks have become cheap and easy enough to use that hackers can now utilize those tools to steal data en masse and then sell it on the black market for a handsome profit.

Now that we know about this game changer, we must ask the question: How should companies protect their commodity data? The answer is this:

through simple economics. If a company can use cyber defenses to change the economics, to make it more expensive to steal their commodity data than it is to steal other companies' data, then the attackers should move on. In other words, make your data the equivalent of an offshore oil deposit when other companies are the equivalent of an onshore deposit.

I have some friends who lived in Alaska for several years. One time they told me that in Alaska, when you go out for a run, you always want to invite someone who is slower than you to tag along. When I asked why this was, they told me, rather dryly, that it is easier to outrun your slow friend than it is to outrun a bear. When your company protects commodity data that other companies also have, you do not have to outrun the bear. You simply need to outrun your buddies.

Pretend for a moment that you are an attacker. You want to obtain credit card numbers that you can sell. To start off your work, you will do some research to select your target and figure out the best way to obtain their credit cards. As you research, you find two potential companies, or victims, that look interesting to you. The first company is a large retailer. Its IT leader was recently quoted in an article explaining that the company just kicked off a large project to improve its security because its current defenses are so outdated. The project just began and will not be in place for at least another year. The quote was intended as an advertisement for the new security technology the company is about to implement. The technology vendor probably gave a generous discount in exchange for the quote. The quote seemed innocent, but to you it screamed "easy target." The second company is also a large retailer. Your research revealed that its IT security leader recently spoke at a large security convention about its cutting-edge cyber defense practices. This security leader also spoke about how they are trying to lead the industry by helping other retailers improve their capabilities. This second retailer is clearly a thought leader in the cyber security space.

Based on this research, which of the two companies will you attack? I'll bet most of us would select the first company. In most cases, people prefer the option that is easier and has a higher likelihood of success.

Understanding all of this will ultimately help businesses that protect commodity information better prioritize their cyber security investment. Cyber security spending is a continuum that stretches from zero to infinity. There is no end to what a company can spend. When looking at this continuum, companies that house commodity data do not have to be on the far high end. They do not have to be cutting edge; they just have to be better than their peers. They have to outrun their buddy, not the bear.

With all the security vendors telling companies how much risk they face and how they should purchase more technology to counter these risks, companies need to prioritize what they really need. Companies that fit the "protecting commodity data" risk profile should realize that their crown jewels are the same as many other companies' crown jewels. Unless a particular company has something that makes its data more unique than the next company, it simply needs to have better security than the next company.

Let's contrast this to a business that houses unique information. This is the equivalent of companies that do not have any friends to invite out when they go for a run. If a bear comes, they are alone and have no choice but to outrun the bear. And, depending on how important that data is, the bear may be highly motivated.

In this situation, the company may also be even more highly motivated to protect the information. I'm not trying to imply that commodity data is not important to protect, but at the end of the day, many retailers have lost quite a bit of customer data and they are still in business. When dealing with unique data, that data may be essential to the future of the company. Many technology companies, for example, exist because they have the best technology.

I used to work in the jet engine industry. It generally takes many years to develop a new jet engine. Large sums of money are spent developing the technology in the engines and then bringing a new engine to market. That money is spent well before the first engine is even sold. If that engine does not sell well, it can cripple the business. It would only take a couple of failed engine launches to put a company out of business.

Now, imagine what would happen if a competitor stole all the technology that is being developed for a new engine. That competitor can go to market with a similar product but at a much lower price point. That competitor can very effectively undercut the business that invested to develop the technology. This only has to happen a couple times before the business doing all the technology investment has to close its doors.

Again, I'm not trying to say that protecting commodity data is not important. I'm trying to point out how critical unique information can be to businesses. Protecting this data may very well be the same as protecting the future of the business.

American Superconductor Corp. is just one example of how losing unique data can cripple a business. In 2013, American Superconductor Corp. was in the middle of a legal battle where American Superconductor accused a Chinese company, Sinovel Wind Group Co., of stealing their software. The American Superconductor software was used to control wind turbines. Sinovel was one of the largest manufacturers of wind turbines in the world and American Superconductor's largest customer.

American Superconductor alleged that Sinovel stole its software and then once Sinovel had the stolen software in its hands, abruptly stopped accepting shipments and cancelled all its orders with American Superconductor. In the wake of the cancelled orders, American Superconductor had to cut almost 60 percent of its employees, going from 842 employees to just 362. The US attorney, John W. Vaudreuil, who sought an indictment, described the incident as "a well-planned attack on an American business by international defendants—nothing short of attempted corporate homicide."[6] Although American Superconductor did not go out of businesses as a result of losing their unique software, the experience was very traumatic for the business and could have easily resulted in a dying company.

In situations like this when the stakes are higher, the level of defense needs to be higher as well. Remember, we are not outrunning our buddy here. We are outrunning the bear, and that bear is very determined. These businesses must have the highest level of defense in order to protect their data and their future.

The concept of businesses that protect commodity data having a different risk profile than businesses that protect proprietary data may sound simple, but there are caveats. The volume of data that a company holds may push a commodity data company closer to a unique data company and increase how determined an attacker will be to break in. For example, let's look at the eBay breach. In 2014, eBay lost the names, addresses, dates of birth and encrypted passwords of 148 million users.[7] Those numbers are huge. The attackers realized that eBay has a lot of what they were looking for and as a result, they were willing to try harder than normal. Did eBay need the highest level of security to defend against this attack? Probably not, but they certainly did have to outrun a couple more of their buddies.

Another caveat that could push a commodity data business' risk toward that of a unique data business is whose information they have. For example, medical records of the President of the United States are more interesting and worth more than the records of an average citizen. When considering where your business falls on the spectrum of commodity data to unique data, it is important to look closely at what you have and think about any uniqueness that may exist in your data.

Also, just as offshore drilling was once perceived to be too expensive and is now common practice, the cyber landscape will also change. Today's defenses that put a company ahead of the pack will someday leave that same company in the dust. Companies that fit this risk profile should set a goal to be better than their peers, regularly review where they stand, and adjust to constantly keep the economics of stealing their customers' data in their favor, not in the attackers' favor.

These two broad categories of businesses that protect commodity and unique data help us understand how risk varies from business to business. They can also help us understand how different businesses should approach their cyber defenses. Let's now take a look at how businesses view cyber security and their cyber security programs. At the heart of every business initiatives should be a clear set of priorities. What are those competing priorities in the cyber security space and how should a business balance them? These will be key decisions that every business must make.

NOTES

1. McKinnon, John D. "Identity Theft Triggers a Surge in Tax Fraud." *The Wall Street Journal*, February 23, 2014, accessed December 26, 2016, http://www.wsj.com/articles/SB10001424052702304834704579401411935878556.

2. Brandeisky, Kara, and Susie Poppick. "How Identity Thieves Stole $5.2 Bil-lionfrom the IRS." Money, September 23, 2014, accessed December 26, 2015, http://time.com/money/3419136/identity-theft-social-security-number-tax-return/.

3. Erb, Kelly Phillips. "TurboTax Temporarily Halts All State E-Filings Amid Fraud Concerns." *Forbes*, February 6, 2015, accessed December 26, 2015, http://www.forbes.com/sites/kellyphillipserb/2015/02/06/turbotax-temporarily-halts-e-filing-in-all-states-amid-fraud-concerns/.

4. "First U.S. Natural Gas Well," accessed December 26, 2015, http://www.lib.umd.edu/epsl/this-week-in-science/first-us-natural-gas-well.

5. "A Brief History of Offshore Oil Drilling," accessed December 26, 2015, http://web.cs.ucdavis.edu/~rogaway/classes/188/materials/bp.pdf.

6. Ailworth, Erin. "Chinese firm charged with stealing tech from Mass. Company." *Boston Globe*, June 27, 2013, accessed December 27, 2015, https://www.bostonglobe.com/business/2013/06/27/feds-charge-chinese-firm-with-stealing-technology-mass-company-amsc/CTE66TzhtD19qvEfU35RQN/story.html.

7. "The 9 Biggest Data Breaches of All Time," accessed June 23, 2016, http://www.huffingtonpost.com/entry/biggest-worst-data-breaches-hacks_us_55d4b5a5e4b07addcb44fd9e.

Chapter 7

Priority Triangle

In previous chapters, we looked at different risk profiles for different businesses. We looked at businesses that store sensitive but nonunique data. These businesses hold information that has value—but that same information is also held by other businesses. We also looked at businesses that house unique data. These businesses are the only ones that have this data. The data may be the designs for the next killer technology, the formula for Coke, or the secret plans for the Death Star. Either way, if someone wants that information, there is only one place he or she can get it.

Think of these two types of businesses as being the anchors at each end of a continuum. Where any particular business falls on that continuum should determine to what extend that business needs the very best in cyber security. In other words, where a company falls on that continuum should impact how that business prioritizes cyber security.

Priority is the greatest challenge that cyber security leaders face today. To better understand priority, let's take a closer look at how businesses can establish and run effective security programs—programs that can address the risk of everything from malicious insiders to cyber doorknob turners. These programs can also address the risk of targeted attackers while also making auditors feel good at the same time.

As I look at security organizations, I believe the biggest challenge they all face is prioritization. How much time and how many resources should organizations apply in order to comply with a standard? How much time and how many resources should companies use to look for attackers who might be hiding in the environment? How much time and how many resources should be applied to closing new security holes as they are discovered? Changing passwords, wiping data off old computers before they are sold off, shutting off access when people no longer need it, the list goes on. In the cyber security space, prioritizing all these potential factors is key.

From my experience, priorities fall into three categories. These categories are compliance, insider threats, and outsider threats. The compliance category includes everything you need in order to comply with your chosen best-practice list or lists and to make your auditor happy. Insider threats include everything you need to do in order to ensure your employees, or the people you trust to access your data, do not abuse that access. Finally, external threats are what we have been focusing on throughout this book. This category includes everything you need to do to detect and respond to all threats from everyday doorknob turners all the way through the most advanced attacker.

In some way, the majority of items that a security program needs to focus on will be covered in one of these three categories. To be fair, a large amount of overlap exists among them. However, when thinking about your security program and how to prioritize your resources, I believe it is helpful to view the space in these three categories. Let's take a closer look at these categories to better understand how you can set clear priorities for a security program.

Think of these categories as three sides of a triangle. If the amount of resources available to any given security program is represented by the area of the triangle, or how much space that triangle takes up, then an adjustment to the length of one side of the triangle will inherently cause a change in the length of the other sides. In other words, if your security program has a fixed set of resources as you increase your focus on external threats, you will inherently have to reduce your focus on either compliance or insider threats.

Some people will argue that this is not true. These people believe the theory that as you become more compliant, you will inherently be better at dealing with external and insider threats at the same time. I understand the theory behind this, but from my experience the theory simply does not play out in practice. Ultimately, you are either judging the success of your security program on whether an audit was passed, whether an attacker was stopped, or whether an insider was prevented from taking something. When you talk about priorities, all three of these items cannot be the first priority. One of them has to be the highest, another one has to be the second highest, and a third has to be last. There is admittedly some interplay and overlap between these areas, but there can only be one top priority.

One of the reasons I like the picture of the triangle is because it captures this interplay between the priorities well. When you lengthen one side of a triangle, it will force you to also lengthen at least one other side while shortening the third. This picture accurately models how these three elements of security overlap. By increasing your focus on external threats, you may get some benefit in the insider or compliance areas, but it will still require you to shift resources away from the core of at least one of those areas.

Security leaders have limited resources, so they regularly face decisions that force them to choose among these three categories. When I work with teams that are fighting off foreign military units and at the same time are being asked to update compliance policy documentation, they have to choose between focusing their time and resources on compliance or on external threats. It is critical that these teams understand the trade-offs they will have to make, and they must clarify which of the three categories is the priority.

Are you still not convinced that we need to make clear priorities in these three areas? When discussing corporate cyber security, it is important to understand that in today's environment, opposing priorities can often drive bad behavior. In chapter 2, we saw how many IT security leaders focus on complying with a predefined list of best practices. The concept is that following these best practices will result in a secure environment. A large part of the IT security industry has accepted this logic. This way of thinking is so ingrained that third parties regularly audit IT security leaders to ensure they are doing everything on their lists. In some cases, such as in the payment card industry (PCI), these auditors have the ability to suspend a business's operations. With the ability to shut a business down, or even preclude it from even bidding on new business, compliance with these lists needs to be a priority for IT security teams. The problem is that these best practices were put together before targeted threats became prevalent. Far worse, people who are not experienced in addressing targeted attacks often maintain these lists.

As we discussed earlier in the book, targeted attacks have changed the landscape of IT security and have ushered in the era of cyber security. Where IT security focuses on best practices as well as on making sure your security program looks as close to one definitive gold standard, cyber security focuses on understanding the attackers, who they are, how they are attacking, and how to stay one step ahead. Where cyber security judges effectiveness by the number of thwarted attacks and how far into the kill chain those attacks got, IT security judges effectiveness based on the opinion of an auditor. This auditor-centric approach will at times oppose an attacker-centric approach. For cyber security leaders, the attacker is the auditor, and you only pass your audit when the attacker gives up and moves to another target.

Throughout my career, one of the things I have learned is that everyone has a boss. Whether it is a boss in the traditional sense or the boss is customers or shareholders, we all have a boss. Whether we realize it or not, how our boss judges our performance will dictate how we approach our jobs. As simple as this concept sounds, I have found it to be very helpful in understanding why people approach their jobs the way they do. After working with auditors for several years, I started to realize that the bosses of a lot of auditors judge their performance based on how many problems they find. This is not true 100 percent of the time, but in the vast majority of situations, an auditor is paid to find issues. If auditors do not find any issues in how a security program is run, then they obviously did not dig deep enough.

This motivation to find issues can often become a challenge for cyber security leaders. As auditors find gaps in a security program, they are motivated to highlight that gap and make it appear to be as big as possible. This is just a natural result of how auditors are judged. If you were paying several hundred dollars an hour for auditors to assess the effectiveness of your security program, and they came back saying they didn't find any issues and everything was good, wouldn't you feel a little ripped off? I can hear it now, "What do you mean?" "We paid a ton of money for someone to find nothing?" The way IT security is set up with its various best practices and auditors testing against those best practices inherently motivates auditors to find issues and portray them in a negative light.

Don't get me wrong, a fine line needs to be walked here, and from my experience the vast majority of auditors do attempt to walk that line. However, just as we discussed earlier, the cyber security leader's job needs to include understanding how the business views risk versus reward. Living in the gray world of accepting certain risks, the auditors need to do the same. Unfortunately, auditors are hardly ever close enough to any given business to do this. Because they lack this business context, their assessments are often skewed, and when they have the ear of senior leaders, those skewed assessments can drive priority conflicts for the cyber security leader.

I recall one instance working with a business that was experiencing a very persistent set of attacks from a foreign military. These attacks were coming in very quickly with attackers changing their tactics multiple times within a single day. The team defending the company was working around the clock catching these attacks and then making adjustments to their defenses in order to prepare for the next wave. The situation was hot and heavy, and all hands were on deck.

While these attacks were coming in, the business was also finishing an audit. During the audit, the auditors examined several of the team's processes. In all cases, the auditor found that the processes where sufficient and operating as they should. However, the auditor did find that even though the correct processes were in place and operating as expected, some of the documentation around those processes was out of date. In other words, the team was doing the right things, but the paperwork that described what they were doing was not up to snuff.

The team was stuck between a rock and a hard place. Auditors were driving them to update documentation; meanwhile, attackers were in the middle of a campaign against them. Imagine that you are on an aircraft carrier that is under attack, and a superior officer orders you to stop defending the ship to drop and give him twenty push-ups. There is a time and place for doing twenty push-ups, but it is not in the middle of a battle. There is also a time and place for updating documentation, but it is not in the middle of an attack. Unfortunately for this team, its auditor did not agree.

Ultimately, the team focused on defending against the attacks and ended up failing the audit in the process. When it was all said and done, the business was kept safe. A foreign military unit was not able to break in, but an auditor deemed the security program insufficient because the Word documents were out of date. Oddly enough, I'm sure the people at the other end of those attacks thought that security program was sufficient. This example highlights that from a cyber security perspective, an attacker is the only true auditor and that the audit is not passed until the attacker gives up.

In all honesty, this situation was extreme. It is not every day that a security team is defending against sustained attacks, and it is not every day that an auditor takes such a narrow view. There is also value in dotting all the "I's" and crossing all the "T's" in a security program. I do not want to diminish the value of audits and compliance. However, I do view adhering to compliance standards and securing a business as two different things. This is why they represent two different sides of our security priority triangle. They are closely related, and to some extent, complying with a standard will certainly make your business more secure. However, complying with a standard does not guarantee security.

This distinction is often overlooked or just outright missed. When the success of your compliance program is judged more by whether or not you have passed an audit than by whether or not you have successfully defended your company, then whether intentionally or not, you have set adhering to compliance at a higher priority than you have set defending against external attackers. If you want to prioritize the prevention of cyber attacks and are doing so by watching your audit results, you are doing yourself a disservice. You will end up driving the wrong behavior.

Look at it this way. When selecting a heart surgeon, do you look for doctors who got all A's on their medical school exams, or do you look for doctors who have successfully completed operations in the past? There is nothing wrong with getting straight A's. In fact, a doctor who got straight A's is probably a very good physician. Regardless, because I prioritize the success of my future operation over the ability to pass a test, I will always prefer the doctor who has successfully completed operations in the past. In the cyber security space, if you want to stop cyber attacks, you need to take the team that has successfully defended against the most advanced attackers over a team that has consistently passed all its audits.

The reverse is also true. If your business is highly regulated and it is imperative that you pass your audit in order to even conduct business, then having a team that successfully stops advanced attackers may not be able to pass an audit. If I need a doctor to help me study for a medical school exam,

I'll want the doctor that got straight A's. Leaders need to know what they are trying to accomplish as their first priority. Knowing this allows you to determine which side of the priority triangle will be the longest. The rest of your security program will fall into place from there.

With all of this said, it is important to note that there are businesses that legitimately have higher priorities than stopping hackers. As we mentioned, if your business is in a highly regulated industry or if your company's ability to do business or bid on new work depends on obtaining a particular certification or passing a specific audit, then compliance should receive more priority. But again, from my experience, compliance does not equal security. I recommend applying enough resources to compliance programs in order to meet your requirements.

When looking at compliance programs, there is always the challenge of selecting which list of best practices you will use and try to comply with. Unfortunately, a number of businesses find that they end up having to comply with not one list of best practices, but several. For example, the payment card industry has a list called PCI DSS. There is also a list from the National Institute of Standards and Technology (NIST) as well as a list from the International Organization for Standardization (ISO) and another list for the Federal Information Security Management Act (FISMA). All of these best practices are known as "frameworks," and these practices are just a sampling of what is out there. Most businesses—either through regulation, customer requirements, or industry standards—need to adhere to multiple frameworks at the same time.

At one point, my security team was responsible for over 1,300 different compliance items, known as "controls," stretched across several different frameworks. At first we spent a large amount of time trying to drive over 1,300 different items. Each of those items required us to put in place a full-fledged business process to make sure everything was done correctly, everything could be audited, and everything would continue to be done correctly in perpetuity. If you have ever implemented new business processes, I'm sure you will relate when I say that implementing over 1,300 processes was not just difficult—it was almost impossible.

As we take a look at our priority triangle, you can see why focusing on compliance will prohibit you from also focusing on external or internal threats. Each of those three areas can be very resource consuming.

From this experience, we learned that there was a lot of overlap across the processes prescribed by each of the frameworks we had to comply with. For example, around thirty different processes needed to be in place to control access to all of our IT systems. These processes spanned everything from ensuring passwords were long enough and complicated enough to making sure access was revoked when people got new jobs and no longer needed access to certain systems. As we examined all these processes, we found that a number of them overlapped. Four of our frameworks included processes for ensuring passwords were at least a certain length. One framework required passwords be at least eight characters long. Another required they be at least fourteen characters long. If we adopted the more stringent of the two and required all passwords to be fourteen characters long, we would effectively meet the two controls. By examining the overlap between the different frameworks and adopting the more stringent of them, we were able to greatly reduce the number of processes we had to put in place.

The second thing we did was to create new computer programs that would manage a number of processes for us. For example, we created a program called an account management tool that managed user accounts and data access. That one program automated all of the 30+ processes that all our frameworks required us to have in place to control access to all our IT systems. By doing this, we greatly simplified what people had to do in order to comply. Once we had this program in place, we no longer had to have people comply with over thirty different processes. We simply had to make sure they used the program we created. If they used that program, we knew all the processes were taken care of.

This program not only took the number of controls we needed to track from 30 to 1, but it also drove adoption of the account management tool. People could choose to use the tool and be left alone, or they could try to do it on their own and have to implement all thirty controls and have them audited on a regular basis. Through this approach, we were able to greatly simplify what everyone had to do in order to comply with our frameworks and pass an audit on them. Rather than implementing over thirty processes for every IT system, we just had to use the one program that already implemented all the processes for us.

By adopting the most stringent requirements from all our frameworks in order to remove duplicate requirements, and by creating programs that automatically handled multiple processes for us, we were able to go from over 1,300 controls to right around 200. Instead of implementing over 1,300 different processes, we only had to implement around 200. As you can imagine, that really made our jobs a lot easier. If compliance is the highest priority in your organization, then I would highly recommend these two approaches as an effective way to simplify and manage all the various requirements you may be trying to address.

The second priority category is insider threats. Throughout this book, we have been examining the risk of hackers breaking into your company. This is admittedly a lopsided view of security. Hackers are not the only risk that security professionals need to worry about. There are also malicious insiders—like Edward Snowden—who have legitimate access to data but choose to abuse that access in order to steal. In 2013 Edward Snowden, working as a contractor for the US government, stole an unknown number of classified documents outlining the United States and some of its allies' surveillance and intelligence-gathering capabilities. Since that time, he has been slowly leaking documents to the press.[1] I'm not trying to comment on the ethical implications of the Snowden leaks. I am using him as an example of someone who had legitimate access to data and abused that access. Regardless of your views on what Snowden did, if he was your employee and leaked your business's data, you probably wouldn't be happy.

When thinking about insider threats, consider both where your business falls on the continuum of commodity and unique data as well as the amount of access people inside the organization have to that data. Understanding these two things can help you prioritize this category. If your business has commodity data and very few people have access to it, then insider threats are not as high of a priority. However, if there is a unique set of data that is the lifeblood of your organization and a number of people have access to it, you should make protecting against insider threats a priority.

If protecting against insider threats is, indeed, a priority, then how can you protect your data against people who legitimately have access to it and work with it every day as part of their jobs? As daunting as this task may be, you can put a couple of principles into place to protect your business against insider threats.

The first principle is to always remember that what you are trying to do is drive the right behavior in the people who have access to your data. A lot of approaches exist for protecting against insider threats. Some approaches require you to identify, tag, and track all of your critical data. This may sound straightforward or easy to do when a sales person from a Data Loss Prevention (DLP) software company is explaining it to you, but believe me, you might as well try to boil the ocean.

Finding all the major IT systems that house your critical data might be easy. For example, if you want to protect the designs of your next product, you will probably find most of that data sitting inside the systems that support your engineering organization. So far so good. However, as you start to dig, you usually find that engineers routinely take data out of those systems and place them in PowerPoint presentations, so they can use them in review meetings. How do you find all those PowerPoint files? You'll also find that your engineers might occasionally email around screen shots of the design as they collaborate with other people. How do you find all of these screen shots? Never mind the files you share with your suppliers, customer, and partners. My point is that any strategy that begins with you identifying and tagging all your critical data is destined to fail. The task is just too big. This is where you need to remember that what you are really trying to do is to drive the right behavior in your employees.

In order to drive behavior, you just need to gather enough data to make a point. I used to use a tool that would track every time an employee touched a file. I didn't care if the file had sensitive data or not. We tracked all of it. Open a file—we tracked it. Print a file—we tracked it. Email a file—yep, we tracked it. Even if someone took a screen shot of a file, we tracked it. As it turns out, tracking access to every file is pretty easy. Identifying only the critical files is hard. Once we figured this out, we started tracking everything.

To make our point, we started using all of that tracking data when people left the company. On exit, we would look for any suspicious file activity. If we found anything suspicious, we would add an extra step to the exit interview and ask if the employee remembered signing a proprietary information agreement that basically said that everything he made while at the company was property of the company. The employee would affirm that he remembered that agreement and had signed it. We would then ask if the employee took anything. If the employee claimed to have left everything, we would let him know that we had a report that showed he had downloaded a large number of files. We would not say what the files were, but that he had a suspicious number of downloads. This would give the employee an opportunity to come clean. If he did admit to taking the data, we would get the data back and all was well. If the employee insisted that the files were family photos or something like that, we would then show him the report which listed not just the number of downloads, but also the names of all the files.

This process established with the employee that he knew his obligation to the company. The employee then had a chance to tell us if he took anything. If he lied, we would present him with some information about the suspicious activity and give him another chance to come clean. If he lied a second time, we then presented him with all the data we had. We intentionally architected the conversation in such a way that it would undermine any legal ground the employee had if we had to go to court. As word of this process got out, it was very effective at driving the right behavior.

As I have worked with various businesses to put this kind of process in place, the worst case I saw was the case of a couple employees who were part of a divestiture. These employees were being let go as part of the divestiture and were getting very generous severance packages in return. As part of their exit, we ran a report and found that they had taken several thousand engineering drawings. Both of these employees persisted in trying to hide their actions. In the end, they were let go and lost their severance packages. Word of this got around, and we could immediately see user behavior change. It doesn't sound nice, but making an example of some people who are acting dishonestly goes a long way to letting others know that you take your intellectual property rights seriously.

Remember, you don't have to classify all your data in order to protect it. You just need to get the people who handle your data to use it appropriately. Let people know that you take the protection of your data seriously and that you expect them to do the same. Communicating this goes a long way.

The second principle is visibility. You'll find that driving behavior addresses most of your insider risk, but there are still those determined individuals who will do whatever they can to steal your information. I've seen employees hired in to a company only to find that they were already the employees of a competitor. Not only were they collecting two paychecks, but they were also passing sensitive data to the competitor. I've seen a husband who worked in marketing and a wife who worked in engineering leave a company only to go directly to a competitor and help them not only make a very similar product, but also position that product the same way in the market. The days of industrial espionage are not behind us.

To deal with these types of insider threats, you will need to do some behavioral analysis on your file access data. You will want to baseline how employees access data and alert when employees begin interacting with data in a way that deviates from their normal behavior. For example, if a person rarely copies files to a removable drive and then suddenly starts copying thousands of files, you might want to look into why all that data is being copied and where that removable drive is located. You will find over time that tracking file access and movement does not just help set up a good exit interview process, but also allows you to detect potential malicious behavior from your employees.

The final side of our security priority triangle is external threats. These threats are the type of attacks that we have been talking about throughout this book. These cyber attacks are the kind that you read about in the newspaper and hear about on the evening news. In chapter 8, we will take a closer look at this side of the priority triangle to better understand how to best prioritize this element of cyber security. We will also examine what you should do if this is the highest priority for your business.

NOTE

1. "Edward Snowden," accessed February 12, 2016, https://en.wikipedia.org/wiki/Edward_Snowden.

Chapter 8

Creating Strong Defenses Against Cyber Attacks

In chapter 7, I introduced the concept of the security priority triangle. This triangle represents where your security program should focus, how many resources that program has, and how those resources should be aligned. We looked at businesses that live inside highly regulated environments and how those regulations drive compliance to the top of the priority list, making it the longest side of a business's priority triangle. We also looked at insider threat, the example of Edward Snowden, and how the people that you trust to work inside your business may be your biggest threat. Now let's look at the last side of the security priority triangle—the external threat. For businesses that are concerned about cyber attacks and wish to prioritize the prevention of them above all else in the security space, let's look at what kind of security program they should put in place.

As people take a look at cyber security programs that address external threats, they may immediately point to the lack of qualified employees in the job market and start to think about how they can outsource this part of their security programs. Based on this, the first question that needs to be answered is "what should be built and run in-house, and what should be outsourced?" The answer for most businesses will be a hybrid between the two, but how should we think about this decision and find the appropriate balance? For a number of businesses, it may make sense to trust components of the security program to a managed security services provider (MSSP) that focuses on security. These providers can often deliver scale and talent that many companies cannot get themselves. For example, to run a security operations center that is staffed 24 hours a day, 7 days a week, and 365 days a year; somewhere between 6 to 12 full-time employees, depending on shift overlaps, are required just to cover the shifts and allow those employees to take little things like vacations and sick days. Most companies do not have enough work to do and cannot keep 6 to 12 people busy investigating all the different alerts that come in. They end up finding that they are paying their analysts to play video games all night because there isn't enough work for them to do. Rather than wasting money on underutilized employees, engaging an MSSP often makes sense.

However, when selecting an MSSP, businesses need to keep a couple of important things in mind. First, scope out what you will outsource to the MSSP and what you will retain in-house. Many people just assume that the MSSP will do everything. This is not the case. Most MSSPs deliver basic analysis capabilities. When an alert fires indicating that something suspicious has happened in the environment, they will investigate and then notify you if they think there really is something strange going on. From there, it is usually your responsibility to do something to fix the problem. In many cases, your team will have to first validate that what the MSSP thinks is suspicious activity truly is, and then your team will have to do something about it. A decent amount of rework is involved. With that said, the MSSP is still usually filtering out thousands of alerts a day, so there is value in what they do, even if you have some rework.

The other thing to keep in mind is that most MSSPs only focus on processing and filtering down alerts. They do not try to connect dots between alerts in order to identify underlying connections that may indicate a broader attack on your environment. Also, when they learn about new malicious activity, they often do not look back in history to see if those new insights reveal anything they may have missed in the past. In other words, the MSSP is simply providing resources to cut down the volume of alerts so that your internal team is not overwhelmed.

I often picture the standard MSSP model as being akin to a beat cop on patrol. Suspicious activity is reported to the police station. The dispatch officer sends a beat cop out to see what is going on. The beat cop is able to verify whether or not something bad is actually taking place and then take steps to stop it. Contrast this with a detective whose job it is to investigate more deeply, to draw connections between different incidents, and to determine if there is a bigger issue. A beat cop is great for stopping people from loitering on corners. A detective is good at determining whether all those people loitering on corners are part of a gang that is trying to move into the neighborhood. Effective law enforcement requires both elements. Effective cyber security also requires both elements. MSSPs are great beat cops, but you still need a detective to watch the bigger picture and make sure nothing is missed.

Security teams are often great at publishing metrics around how many attacks they stopped in the last week or month. The numbers are always large. What a lot of people do not look at is how many of those attacks were related and coming from the same set of attackers—a set of attackers that were actively prodding their defenses trying to find a hole. Stopping 1,000 attacks a week is one thing. It is something else to know that five of those attacks were not random, but that they were targeted and part of a larger effort to compromise your organization. If you do not have the capability to connect these dots and understand when attacks are connected and targeted, then next week you may only stop 999 attacks instead of 1,000 because the attackers finally found a way around your defenses, and you are none the wiser.

This type of analysis, or investigation, is what a detective performs over and above a beat cop. Your normal MSSP does not provide this type of analysis or investigation. If you are looking for an outsource partner, keep this in mind. Hiring beat cops is good and valuable, but you still need some detectives. With that said, regardless of your approach to outsourced services, you still need to make some fundamental choices when thinking about a corporate security program. MSSPs are often very good at handling the door-knob turning type of attacks; they cover some compliance requirements as well, but they traditionally do not help very much with targeted attacks. Most MSSPs are not set up to deliver the kind of custom detection and support that is needed to properly defend against targeted attacks. Keep this in mind as you look at your outsource model.

At the end of the day, there is a difference between a random crime and one that is targeted and focused on a specific victim. Let's imagine for a minute that you are the proud owner of a very exclusive and expensive exotic car and that you keep this car safely inside your garage. You are obviously very proud of your car because it is the only one of its kind in the entire state. Now imagine that you wake up one morning to discover evidence that during the night someone has unsuccessfully tried to break into your garage. What is the first thing that runs through your mind? Most of us would assume that the thief was attempting to steal the car. We would think that there is someone out there who is trying to get your car. Now, how would your feelings change if, as the police were writing their report, they told you that every house on the block had the same issue? Wouldn't you feel like your car is somehow safer? It now appears as though the thief isn't after your car. It looks as though you and your neighbors were the victims of a random crime. It no longer looks as though you and your car were specifically targeted. As you can see, there is definitely a difference between random crime and targeted crime.

If your garage were the only one that the thieves tried to break into, wouldn't you be worried that they will come back? If your garage were the only one targeted, wouldn't you be looking at ways to improve the security of your garage? Compare this to the steps you would take if everyone in the neighborhood were hit. Would you be as worried? You might make sure your doors are locked, but you probably wouldn't be installing a new security system. When you are dealing with random attacks, the risk just is not as high as when you are dealing with a targeted attack.

For any business leaders who are concerned about protecting their crown jewels and who make sure they take every step necessary to do so, there are proven approaches. However, these approaches do not focus on security controls. They do focus on getting visibility into what is taking place on computers and on computer networks. This visibility not only allows you to detect attacks, but it should also help you determine whether the attack you just detected was random or targeted. Just like our example of protecting an exotic car, understanding whether an attack was random or targeted is critical when determining how concerned you should be.

Going back to our car example, imagine now that your garage was the only one that the thief attempted to break into. A week has passed, and you wake up to find another failed attempt to break into your garage. Now your garage has been targeted twice. How would you feel now? What measures would you take? This second attempt now confirms that someone is after something in *your* garage, and it appears as though he will keep trying to get in there until he is either successful or apprehended.

This scenario highlights two things. First, the more you are targeted, the higher your risk will be. Second, the more you are targeted, the more you will need to do to protect yourself. In this situation, you were probably fine when a beat cop showed up after the first break-in attempt and wrote a standard report. After the second attempt, you will probably want a case opened and a detective assigned to it. If these attempts to break into your garage keep happening, you don't want to be talking to different beat cops each time, you want to deal with one detective who is able to look across all the incidents, make connections between them, and examine other incidents around town or even across the state or country. You want someone that has visibility to information beyond the single report on the last break-in attempt.

These two concepts hold true in cyber security. The more you are targeted, the higher your risk and the more you will have to do. Step 1 is figuring out if the attacks you face are random or targeted. Step 2 is determining if the attackers are continuing to come back, trying again and again to break in. If they are, then you really need to up your game. Relying on technology that does the same thing over and over will not work, as attackers figure out how to circumvent that technology. Relying on a standard outsourced MSSP that is focused only on the report for the latest break-in attempt will not work either. The MSSP is the beat cop. Each incident may have a different analyst looking at it, and they are not looking beyond that single incident. You need the cyber equivalent of a detective to investigate more broadly to draw connections between the attacks you are seeing and attacks that other companies are seeing to help you better understand how best to adjust and change your defenses as the attackers change their attack. If your security organization is only telling you about how many attacks it stopped and not how many of those attacks were from the same bad guy, then it is safe to assume your security team lacks someone in the cyber detective role and that they do not have the ability to consistently and effectively determine whether your company is the victim of random attacks or targeted attacks.

Another point to keep in mind when dealing with targeted attacks is that having the latest technology is not sufficient for protecting the businesses. You have to have the ability to constantly morph and change your defenses. Like the dinosaurs in *Jurassic Park*, systematically trying different parts of the fence searching to find a weak spot to break through, attackers will continually probe your defenses. Over time, they will understand how you defend yourself and craft an attack specifically designed to evade your defenses.

The only way to protect yourself is to create a defensive posture that continually morphs and changes. When you do this, attackers cannot get a clear picture of your defenses. Each time they come back, they are faced with different challenges.

Think about an attack from an attackers' perspective. As we saw in the Kill Chain, there are numerous steps attackers have to go through before they have a successful attack. In fact many times, just like with the Sony Pictures Entertainment breach, attackers may be operating inside their victims' computers for months before they achieve their goals. From an attacker's perspective, if everything goes as planned, the attackers will get their backdoor installed, and they will obtain command and control of a computer inside the victim's network. From that compromised computer, they will figure out where they are and where the information they want is. They will have to explore the computers to figure out the lay of the land, where things are stored, what kind of access they need, and so on.

An attempted attack is a bit like being placed in the middle of a maze you have never seen before and told to find the pot of gold hidden somewhere in the maze. The challenge is fun, but it takes time. All along the way there are booby traps—AKA security defense—trying to stop you. You need to identify what these defenses are and work around them. Finding the pot of gold doesn't sound very easy, does it? Truth is, executing these attacks is not easy, but enough of them have been pulled off that techniques and tools for walking through these steps have been developed, and as these techniques and tools evolve, attackers are better able to navigate these proverbial mazes and find their pot of gold.

Now, imagine that attackers are inside the maze, trying to find the pot of gold, but they discover that the maze reconfigures itself on a regular basis. Just when they think they have it figured out, the maze changes and the game is entirely new. Companies trying to protect unique information need to put these kinds of defenses in place. Such defenses are difficult to implement, but they dramatically raise the difficulty level for attackers.

Many approaches to creating morphing defense exist, many of them fancier and more complicated than the next. From my experience, a fairly straightforward approach exists for creating morphing defenses. This approach is based on the fact that attackers use tools to execute their attacks and help them explore and understand the maze they are trying to get through. What many defenders do not understand is that when they catch an attack, what they actually caught is one of the attacker's tools, and these tools can be used to track attacker activity.

Often attackers will take a tool out of their toolbox and attempt to use it on a victim. If the tool doesn't work, they will adjust the tool and try it again. This process often continues as they keep adjusting their tools, trying to poke at different parts of a company's defenses. While attackers are adjusting their tools, defenders can learn from these tools. Once they know how to catch the tool, they can observe how the tool is changing. As the tools change, the defender should also be changing their defense. It is a bit like the *Jurassic Park* staff, dynamically extending the fence based on where the dinosaurs are roaming. By implementing a process where defenders detect attacks, learn from those attacks, and then morph their defenses based on that learning, defenders can create dynamic defensive postures that are designed to stop the specific attackers they are seeing.

This may sound like a lot of work, and it is, but it is the most effective way to keep determined attackers out of computer systems. Only morphing defenses that constantly move can stop attackers from understanding what they are up against and effectively defend against the most sophisticated attacks.

When prioritizing external threats, consider where your business falls on the continuum between having commodity data and unique data. We talked about this concept at length earlier. If your business is protecting commodity data, then you can shorten the length of the external threat side of your triangle. You do not have to have the best protection against external threats; you just need to be better than most. However, if you are protecting unique data, then consider extending the external threats side of your triangle. Protecting unique data, data that no other company or organization has, means external attackers will be more motivated and willing to go to greater lengths to obtain your data.

If external threats are your priority, then you should focus on implementing morphing defenses in line with what we discussed earlier. Creating morphing defensive postures that present an ever-changing set of challenges to attackers is the single best way to counter high-end external attackers. We talked about catching and tracking the tools that attackers use, but what else needs to be in place in order to create dynamic defenses?

There has been some noise in the security industry around creating morphing defenses with networks and other computer systems that continually move and reconfigure themselves. Let me just say that as a user of IT systems, I already struggle to find what I need. I can't imagine how much harder it would be to do your job if the systems kept moving. This kind of dynamic environment doesn't just make the attackers' lives harder. It also makes your employees' lives harder. Cyber security professionals need to find ways to inflict maximum impact on attackers with the least amount of impact on corporate users.

Morphing defenses should exhibit two characteristics. First, morphing defenses should make it hard for attackers to determine how they were caught. From an attacker's perspective, it is very hard to avoid a defense that you cannot see; whereas, if you know what is stopping you, you can figure out a way around it.

When I was growing up, my father worked on cars. To this day, he has a constant stream of cars that he restores going through his shop. As we worked on these various cars, I became pretty proficient in putting bolts on and off. I learned that pretty much any bolt can be removed. Whether it was rusted in place, completely seized, stripped of threads, or even if the head was torn off, it didn't matter. We could still remove the bolt. The one exception to this rule is removing a bolt that I couldn't see. There is nothing more frustrating than trying to take out a bolt I cannot see. Do I have the right size wrench? Is the wrench on the bolt properly? Is the bolt rusted into place? I don't know—I can't see it. When you can observe what you are up against, you can take measures to get around it. If you cannot observe what is hindering you, then you not only have a much more difficult task ahead of you, but you also have a much more frustrating task ahead of you. Likewise, anything that frustrates the attacker is good for the defender.

Let's consider most of the current security technology on the market. Firewalls or what is popular now—next generation firewalls—are an essential part of any security program. Firewalls essentially permit or deny communications between computers based on specific criteria. If your communication meets the criteria, the firewall lets the communication through. If your communication does not meet the criteria, the firewall stops that communication.

From the attackers' perspective, when their communications are stopped, they immediately know they have a problem. Now, they may not know that a firewall is blocking the communication, but the attackers do have real-time feedback that something is out there preventing their communication from going through. Whatever it is that is stopping their communication is effectively stopping their attack. The attackers know they are stopped, and they need to figure out a way to make their communication work. As the attackers adjust how they are communicating, they will get real-time results to know whether their adjustment worked or not. If their adjustment allows the communication to meet the firewall's criteria, then the communication goes through and the attack continues. If the adjustment did not meet the criteria, then the attacker knows that immediately and continues making adjustments until something works. In short, the attacker is seeing the defense and can interact with the defense, poking and prodding it until the attacker finds a way around.

Compare the firewall to a dynamic defense that obscures how an attack is detected. With dynamic defenses, the defender is alerted to the attacker's presence. At this point, the defender determines when to step in and stop the attack. Recall the steps in the Kill Chain. Let's assume that a defender caught the delivery of the backdoor. The defender knows that an attacker has sent a backdoor into the environment. At this point, the defender can either block the installation of the backdoor, or let the attack progress and wait to stop it at a later phase. This sounds counterintuitive. Why would someone let an attacker progress with his or her plan?

This game is one of cat and mouse. If a defender stops the attack when the backdoor is delivered, the attacker will know that the defender can successfully detect that backdoor. In other words, the attackers know they were caught when they sent that particular backdoor. If your attackers are any good, they won't send that backdoor your way again. The next time, they may send a backdoor you can't catch. However, if you let the attack progress, the attackers will assume that their backdoor slid by your defenses and worked as they wanted it to. This increases the odds that when the attackers come back later, they will reuse that same backdoor that you already know how to catch. By obscuring from the attackers how they were detected, you place them in a very uncomfortable position. Without knowing how they were caught, they have no idea what they can reuse from attack to attack. This means you can track them by the way they reuse various tools across their attacks, or if they choose not to reuse any tools, then you have very effectively increased the cost for the attackers because you have forced them to make new tools. Obscuring how your defenses detect attacks is important, whereas any technology on the market that automatically blocks attacks does not obscure anything.

The second principle for dynamic defenses is the ability to rapidly change how you detect attacks. Every couple of years, the security industry comes up with a new way to catch attackers. Several years ago, behavior-based detection was all the rage. The concept behind behavior-based detection is that if you open a file and watch what it does, that file will show you whether it is good or bad based on how it behaves. For example, if you open a PDF document and that document tries to change your system settings, then chances are you have something malicious in the PDF because PDF files do not normally edit system settings.

Before this approach was conceived, most detection was done using signatures. The problem with a signature is that you generally need to know what you are trying to catch before you can catch it. If someone writes a new malicious program that you have never seen, there is a good chance your signatures will not catch it. Based on this new behavior-based detection concept, you no longer needed signatures. You could effectively catch malicious programs you have never seen before just by the way they behaved. It was very effective and the security market was flooded with products that would open files, watch what they did, and report any suspicious behavior.

Several years have passed since behavior-based detection was billed as the silver bullet. Attackers now have malware that allows them to detect when they are being observed and therefore turn off malicious activity. It didn't take attackers long to figure out how they were being caught and figure out a way to get around it. As this happened, the security industry has recently come forward with a new approach based on analytics. The concept is that by watching a very large number of data points, attackers will statistically stand out from your normal users.

Regardless of the merits of this approach, I believe it highlights the need for defenses to have multiple ways to detect attackers. Attackers figured out how to get around signatures by creating new malicious programs for each attack. Attackers figured out how to get around behavior-based detection by turning off their bad behaviors at the right time. Attackers will figure out how to get around analytics-based detection someday as well. Because of this game of cat and mouse, defenders need to have multiple detection techniques at their disposal.

I have found that the more attackers try to hide from one way of being detected, the easier it is to detect them via another method. Defenders must also have the ability to shift between each of these techniques very quickly as attackers change their attacks. If your business needs to make defending against external threats a priority, then creating dynamic defenses with obscured and rapidly changing detect capabilities is critical.

If you are able to both obscure from the attackers and rapidly change how you detect attacks, you will have the basic building blocks required to create dynamic morphing defenses. Couple this capability with the equivalent of a cyber detective to coordinate how attacks are stopped, as well as how detection is changed over time, and you will have a very effective security program.

But what about all that technology out there that actively blocks attacks? What do we do with that? Remember, there are two kinds of attacks: random attacks and targeted attacks. Active blocking technology is perfect for random attacks. These are the equivalent of cyber doorknob turners, people who walk the Internet trying doorknobs looking for an unlocked door that they can walk through. Addressing these kinds of attacks is a perfect application for tools that automatically block. If the attack is stopped, the attacker will move to the next target, just like someone trying doorknobs. If the door is locked, they move on. There is no game of cat and mouse; there is no attacker coming back again and again trying new ways to break in. The door is locked, they move on. Automated defenses are perfect for this and quite frankly when you look at the volume of "attacks," the vast majority of them fall into this category. However, the biggest risk comes from those few attacks that are targeted at you and your business.

We need to look at external attacks in their entirety and use automated defenses to protect against the random attacks and use dynamic morphing defenses to protect against the targeted attacks. This one-two combination will be very effective against all forms of external attacks. As we look back at our security priority triangle, I hope you can see the difference in the complexion of a security program focused on external attacks versus the one focused on compliance versus the one focused on insider threats. Where external threats drive a program to implement dynamic morphing defense, internal threats drive a program to drive employee behavior and gain visibility in to how employees handle data.

Some overlap and shared benefit exist, but the focus is different. While compliance-focused security programs will zero in on different security controls and driving those throughout the entire IT environment, external threat-focused programs will look more at who is attacking an IT environment and how those attacks are changing. Again, there is some overlap and shared benefits, but the focus is different. As you set priorities for your cyber security program, think about this priority triangle, and make sure you are clear which of these areas needs to be the primary focus for the business. These concepts will go a long way in helping to establish a strong corporate cyber security program.

Let's shift gears and take a look at personal security. If corporate security is this complicated, what are the risks we all face as individuals and what can we do to protect ourselves? We will examine these questions in the next chapters.

Chapter 9

Personal Security—How Big Are the Risks?

We have examined the issues and risks associated with cyber security within the corporate realm, but what are the risks to each of us as individuals? Most of us do not have millions of credit card numbers, medical records, or designs for the next must-have product like many companies have. Why would we be the targets of a cyber attack, and just how big are the risks that we face every day as we go about our ever increasingly connected lives?

It never occurred to me just how much I rely on the Internet and the information it provides. Recently my wife and I spent some time in a cabin located in a fairly remote area. Our first clue that we may had misjudged the situation came as we were driving to the cabin. My wife called a local restaurant to make reservations for dinner. After scheduling our time, the reservationist asked if we needed directions. My wife declined, telling him that we will just use our smartphones to find the restaurant. The person on the other end of the phone had to politely explain that there was absolutely no cell coverage in the area and that we had better take him up on his offer for directions.

We then checked into our cabin and received an information sheet for where we would be staying. At the bottom of the page was a place for the Wi-Fi password. On our sheet, the password was listed at "NA." When I saw this, I thought, "Wait a minute, that's not a valid password." It was then that I realized that the password wasn't written down wrong, but that we simply didn't have Wi-Fi. It took me a couple of minutes to stop twitching.

After figuring out what we had gotten ourselves into, we realized we did not properly prepare for this. We expected to get there and do a Google search for restaurants, for things to do, and for places to go. Without any connectivity, we managed to get lost as we tried to drive to the nearest town so we could get back online and plan the rest of our trip.

Whether it is planning a trip, finding a place to eat, or figuring out how to get from point A to point B, I have become very accustomed to using the Internet, and I would suspect that I'm not alone in this. We are connected people who live connected lives. This connectivity allows us to get more done, avoid traffic, find the best places to eat, and generally live better lives. But just as we saw in the business world, with all this connectivity comes risk.

We have already looked at the risk of stolen credit cards and other personal data. In the case of credit cards, the risk is relatively low because most credit card companies do not hold the card owner responsible for fraudulent charges. In the case of stolen credit card numbers, the issue is more an inconvenience as you switch out your old card for a new one.

Stolen personal data, on the other hand, can be more serious. As we discussed earlier, stolen personal data can lead to identity theft that can result in a number of issues. If an attacker opens new accounts under your name, creating new credit cards or other lines of credit, or even filing fake tax returns, the results are a lot harder to clean up than just switching from one credit card to another. In many cases, identity theft can continue as criminals keep using your information to cause new trouble for you potentially years after they originally stole your information.

The challenge with identity theft is that it is the gift that keeps on giving. Your mother's maiden name really doesn't change. Your Social Security number doesn't generally change either. The information that attackers steal in order to perpetrate identity theft is generally information that does not change. Because of this, unless you take steps to freeze your credit or further protect your identity, attackers can continue to reuse this stolen data even years after they originally got their greedy little hands on it. In chapter 10, we will explore how we can protect ourselves from identity theft.

When thinking about personal cyber risk, there are more things to think about and consider than credit cards and identities. People are also trying to gain access to your online banking account. Over the last couple of years, there have been an increasing number of malicious programs known as "banking trojans." These nasty pieces of software install themselves on your computer and then wait for you to log into your bank account. When you log in, they attempt to steal your user ID and password and send it back to the bad guys who can use that information to transfer money out of your account and into theirs.

One of the biggest and most popular of these banking trojans is named "Zeus" after the king of the Greek Gods. As it turns out this name is very appropriate. Zeus was first seen in 2007. Since then, it has infected tens of millions of computers around the world and has been used to steal hundreds of millions of dollars. It was so popular that the creators of Zeus eventually followed the model of commercial software companies by licensing the right to use the program to other bad guys. In the same way you can buy a license to install a copy of Microsoft Office on your computer, cyber criminals would purchase a license to install Zeus on other people's computers in order to steal their banking user ID and password.

Eventually the source code—all the programming that made Zeus work—was released on the Internet. This was kind of like publishing the secret sauce that made Zeus so effective. Since then, Zeus has continued to cause trouble, but it is not the king god that it used to be.[1] There are now numerous copycat programs out there that help bad guys steal your banking user ID and password. Once they have this information, they are able to transfer funds out of your account.

But wait just a minute, you might be saying. How can these guys just transfer money out of my account and not get caught? Can't the authorities just trace the money and find the owner of the destination account? With that information, they should be able to arrest the bad guys. That's a good point, and that is why many of these schemes involve what is called "money mules." A money mule is an intermediary that essentially launders the money for the bad guy. The entire purpose of having a money mule is to make it harder for the authorities to track the bad guy. What is so shocking about this system is that most money mules do not even know they are part of a fraud.

Have you ever seen one of those ads on the Internet stating that a stay at home mom is making some obscene amount of money working part time in her home? These ads usually say that you can do that same thing. In a number of cases, the ads are tied to money mule schemes. The innocent stay at home mom is interested in making some money on the side so she can help pay bills, but she still needs to be home with young kids. This ad looks like a perfect solution.

The scam goes something like this. The person looking to make some extra money (the mule) responds to the ad and is told that a large company needs help moving money between its various accounts. To do this, the company will transfer money into the mule's account. All the mule has to do is then transfer that money into a foreign account using a financial services company. In exchange for helping this company transfer its funds, the mule can keep a certain percentage.

From the mule's perspective, this job may sound a little bit strange, but when the first deposit is made *into* her account, her fears are often put to rest. When you think about it, most of us are watching to make sure we are not being taken advantage of, or conned out of something. In this case, we would be concerned that money would be removed from our accounts. However, once we see money coming into our account, many of us would begin to feel comfortable. This is exactly the way the perpetrators of this scheme want their potential mule to feel. "This is not a scam. I am working with a company helping its accounting team perform an important function, and I get to make pretty good money in the process." What could be better?

Unfortunately, the authorities eventually come knocking on the money mule's door. At this point, the mule finds out that she has been participating in and enabling a fraud. The money mule is a very effective way for these cyber criminals to launder their money and essentially throw the authorities off their scent. Unfortunately for the money mule, she finds out that she has been unwittingly breaking the law and can be prosecuted for it.[2] Too bad that ad on the Internet didn't tell her about that part.

Beyond financially motivated risks, are there other things that we need to watch out for or at least be aware of when it comes to personal cyber security? The answer is yes. As more and more of the things we buy become "Internet enabled," or have the ability to connect to the Internet, we are seeing and will continue to see additional risks.

There is a very popular term in the technology space called the "Internet of Things." This is a broad term that refers to the trend of everyday items connecting to the Internet. It started with your television connecting to the Internet, which allows you to watch Netflix or Hulu. From there, your thermostat connects to the Internet so you can set the temperature of your home while you are away. The trend of connecting things to the Internet has only picked up speed from there. The Internet of Things has already grown quite large. Recently, Mattel released a Barbie doll that is able to connect to the Internet. This Barbie can listen to your children, and via the power of the Internet, can interact with them. The idea is to have a Barbie doll that can interact with your children similarly to how Apple's Siri interacts with you.

All this connectivity is great and brings additional convenience to our lives, but at what cost? Take that Barbie doll, for example. After it was released, it didn't take long for researchers to figure out how to hack into the doll and gain access to much of the information it contained, including the recorded audio of the children who used it. The researchers claim that it is now only a matter of time before they figure out how to configure the doll to talk to their servers instead of the Mattel servers. Once they do this, they can not only listen to all the audio stored on the doll, but they can also control how the doll responds. They can essentially make the doll say whatever they want.[3] How creepy is it to think that someone could use your child's Barbie doll to not just listen to you, but to also talk to you and your children? It's almost like those old horror movies where the doll comes to life.

Unfortunately, this invasion of privacy does not stop with Barbie dolls. Researchers at Rapid7, an IT security company, conducted a study to find out just how secure Internet-connected baby monitors are. Many parents love these devices because they can watch what is happening in their baby's room via their computer or even their smartphone. The camera essentially makes full audio and video feeds and then sends those feeds via the Internet to a website or an app on a smartphone where the parents can access them. Some of the higher-end models even have speakers built in so that parents can remotely talk to their children.

While researching these devices, Rapid7 found that a large number of them were relatively easy to break into and control. Some devices had passwords that could be used to gain access to the devices. This in and of itself is not a major issue, but a number of these passwords were unchangeable and either printed in the user's manual or freely available on the Internet. With a little research anyone in the world could gain access to some of these cameras.[4] This is just another example of a product that has not been traditionally connected to the Internet inadvertently introducing new risks to its users as it adds new features. I'm sure the designers of the Barbie doll and the baby monitors did not intend for their products to be turned into spying devices. They probably just didn't know the risks that were out there and didn't take appropriate measures to address those risks. As the Internet of Things continues to proliferate, this problem will become more and more common. Now all of a sudden, engineers who have zero cyber security experience are designing products that expose their users to cyber security risks. I believe that over time, this will lead to a certification or seal of approval process where, similar to how the Underwriters Laboratories tests electronic devices for safety, someone will eventually test all "connected" devices for cyber safety. In fact, Underwriters Laboratories just announced that it is launching just such a program.[5]

In addition to the potential invasion of privacy, the Internet of Things also highlights the weakest link issue. The "weakest link" is a term used to convey the fact that the security of an overall system is only as good as its weakest link. In the case of the Internet of Things, not only could someone hack into your baby monitor to watch and listen to what is happening, he or she could also use that monitor as a way into your home network, potentially using the monitor as a launching point for an attack on your computer or any other Internet-connected device in your house. Yes, having more and more products with the ability to connect to the Internet will bring more convenience into our lives, but until the engineers developing those products understand all the cyber risks, we will continue to see these technologies exploited.

If you still aren't concerned, I have one more personal risk to point out. This is the risk of cyberstalking. Cyberstalking is exactly what you think it is. Essentially stalkers use the Internet to follow, harass, and intimidate their victims. What makes cyberstalking so difficult to stop is that unlike physical stalking, it can be done from the other side of the world. With international

law as it is, many countries do not have laws on the books against cyberstalking, and even if they do, getting the authorities to prioritize such a case when
they already have more than full dockets is often difficult.

Newsweek recently published one well-documented instance of cyberstalking. The story is about a mezzo-soprano singer named Leandra Ramm who
had a blossoming music career. Leandra was contacted by Colin Mark Yew
Loong, who was posing as the director of a music festival. Colin offered to
help Leandra with her career, but it didn't take Leandra long to figure out that
Colin was not who he said he was and stopped responding to his messages.
Events then took a turn for the worse.

In typical stalker fashion, Colin began sending threatening emails.
He used the power of social media to try to discredit her by making groups
on Facebook and Twitter that published false negative information about
Leandra and her career. Colin created a blog where he made rape and other
physical threats. He even went as far as to make bomb threats against organizations that hired her.

All this cyberstalking eventually ruined Leandra's career and sent her into
a dark time where she was diagnosed with posttraumatic stress disorder and
even thought about suicide. To make matters worse, despite having proof of
all the emails, blogs, and Facebook and Twitter posts, Leandra was unable to
see Colin brought to justice. Colin was in Singapore, and Singapore did not
have any laws against cyberstalking.

Eventually Leandra was able to work with cybercrime experts, and she
was able to get the help she needed to work through both the American and
Singaporean justice systems. Colin was finally brought to justice. In the court
filings, two other women and one man were also named as victims. Leandra
was not Colin's only victim. His other victims included a Ukrainian musician
and a Singaporean businesswoman as well as the boyfriend of a Hungarian
musician. To these other victims he sent harassing messages, made death
threats to them and some of their children, and even threatened to castrate the
boyfriend of the Hungarian musician.

In the end, Colin pleaded guilty to his charges and was jailed for three
years. Matthew Joseph, the district judge, called this case an "abhorrent
case of cross-border cyberstalking." When he spoke to Colin, the judge said,
"The virtual Internet in your criminal hands became a lethal weapon. It was
used as a weapon of massive personal destruction in the real world of your
hapless victims." This case was the first successfully prosecuted international
cyberstalking case in history.[6] Let us hope that justice continues to be served
in these situations.

Whether we are talking about financial risks, privacy risks, or even personal
risks, cyber security affects not just companies but each of us as individuals.
Just as my wife and I realized how much we rely on the Internet when we

suddenly found ourselves without it, all of us use the Internet to make our lives better. As with everything, there is a good and a bad side. By highlighting some of the personal risks in the cyber world, I am not trying to say that we should all pull the plug, drop off the grid, and wear tinfoil hats. I am simply trying to highlight the risks that we all face every day.

Just as driving a car introduces the risk of a crash, using the Internet introduces the risks we just discussed. We learn how to drive safely, and we wear seatbelts in cars to reduce the risk associated with driving. In chapter 10, we will learn how to protect ourselves in the cyber world.

NOTES

1. "The Big Four Banking Trojans," October 21, 2013, accessed February 27, 2016, https://blog.kaspersky.com/the-big-four-banking-trojans/2956/.

2. "Money Mules," accessed February 27, 2016, https://www.safeinternetbanking.be/en/fraud-techniques/money-mules.

3. Gibbs, Samuel. "Hackers Can Hijack Wi-Fi Hello Barbie to spy on our children." *The Guardian*, November 26, 2015, accessed February 27, 2016, http://www.theguardian.com/technology/2015/nov/26/hackers-can-hijack-wi-fi-hello-barbie-to-spy-on-your-children.

4. Stanislave, Mark, and Tod Beardsley. "HACKING IoT: A Case Study on Baby Monitor Exposures and Vulnerabilities." September 29, 2015, https://www.rapid7.com/docs/Hacking-IoT-A-Case-Study-on-Baby-Monitor-Exposures-and-Vulnerabilities.pdf.

5. Higgins, Kelly Jackson. "Underwriters Laboratories to Launch Cyber Security Certification Program," accessed April 6, 2016, http://www.darkreading.com/endpoint/underwriters-laboratories-to-launch-cyber-security-certification-program/d/d-id/1321202.

6. Quarmby, Katharine. "How the Law Is Standing Up to Cyberstalking." *Newsweek*, August 13, 2014, accessed February 26, 2016, http://www.newsweek.com/2014/08/22/how-law-standing-cyberstalking-264251.html.

Chapter 10

Personal Approach—What Individuals Can Do to Protect Themselves

There seems to be an unlimited supply of cyber ne'er-do-wells out to take advantage of anyone who lets his or her guard down. These cyber space predators don't really care who their victims are. They lurk around the Internet looking for anyone to pounce on. What can we as individuals do to protect ourselves in this brave new world?

A number of the challenges to our personal security in cyber space stem from two things that the Internet provides: anonymity and obscurity. Because of the inherently impersonal nature of the Internet and, as we saw earlier, the fact that basic security mechanisms were not built into the Internet as it was made, truly knowing who you are communicating with is extremely difficult. You may be thinking, "Wait a minute." How can this be? You regularly communicate with coworkers, friends, and family on a daily basis. Whether via email, chat, text, or any other technology, you constantly communicate with people you know. But, do you *really* know who is at the other end?

Take, for example, a scam that was quite popular a couple years ago that involved an attacker breaking into a person's Facebook account and then using that account to send messages to the victims' friends asking for money. There would be an elaborate cover story, generally around the owner of the Facebook account being on international travel and a series of disasters that supposedly left the person stranded and without enough cash to make it back home. Does this sound familiar? It usually didn't take long for the Facebook account owner to realize that someone had taken over his account. Once the account owner realized what was happening, he generally sent a message to all his friends letting them know that the message asking for money did not come from him.

You may still see this scam being executed today. The Facebook account owner usually says something along the lines of "I'm sorry, but my Facebook account was taken over. You may have received a message from me asking for money. Please ignore the message and do not send any money. I have let Facebook know about this." By the time this message comes out, someone may already have sent money and given the attacker a payday.

At the root of this scam is the fact that you cannot truly verify whom you are communicating with on the Internet. So many of us take what we see at face value. Just like Leandra Ramm (see chapter 9), who assumed the person she was talking to was a director of music who represented a music festival and wanted to help her career, many of us assume what we are told is true. Most of the time this assumption is valid, but it is good to have a healthy paranoia when we receive unsolicited communications via the Internet.

This means that we have to be especially careful to verify whom we are communicating with before we send money, personal information, or anything we wouldn't want published on the front page of a newspaper. In fact, the "newspaper test" is a great rule of thumb to use. As you communicate with people in cyber space, you should ask yourself: "Am I absolutely sure I know who I'm communicating with and that they are who they say they are?" If not, only send information that you wouldn't mind seeing on the front page of a newspaper.

Once information leaves our control on the Internet, it tends to take on a life of its own. Because the Internet makes it so easy to transmit and store information, that information is regularly copied and stored in numerous places without us even knowing. For example, when you take photos and store them in iCloud, your photos are stored on your phone, on Apple servers, and potentially across any other device you have configured with your photo stream. All of this happens automatically without you doing a thing.

In a previous job, I was responsible for the enterprise data backup strategy for a global company. As we looked to reduce the amount of storage we used, we ran a project to look at how many copies we kept of any given file. Between various backups schedules, replication schemes designed to make sure data was still available even if a hard drive completely crashed, we found that for any given file we actively maintained well over ten copies of it. Some of those copies were sent to permanent off-site storage just in case they were ever needed way down the road. I can guarantee you that the people who made those files never expected so many copies of them to be made and that they would be kept for so long. Whether we think about it or not, digital is forever.

When you make hand-written notes on paper or even hold old school photos in your hands, you always have the option of destroying those notes or photos. Did you ever write a letter that you now find embarrassing? Did you ever take a picture that you wish you hadn't taken? Don't worry—a little bit of fire will destroy those paper documents.

In cyber space, that's not the case. When your information is digital, it takes on a life of its own. Not only are files copied without you knowing it, but also notes and messages that you originally intended for one person could get sent to people you never meant them to go to.

Recently I used the search function on my smartphone and found notes from years ago show up in the search results. I thought those notes disappeared years ago. Heck, I didn't even have my phone when I wrote those notes. Notes are just the tip of the iceberg. Pictures . . . well just look at how many people have gotten into trouble or have been embarrassed by pictures they wish they never took. The same is true for posts on social media. Did you have a bad day years ago and let off a little steam on social media? A future employer can potentially find that post and get the wrong impression of you. Digital is forever, so apply the newspaper test the next time you communicate with someone on the Internet or store something in the Cloud. Not only does data tend to live on forever, but in a lot of cases, the information you put out there can potentially reach a much larger audience than you think.

Imagine for a moment that you have the opportunity to give a short speech that will be broadcast around the world. People in almost every country will hear your speech. You have just been given an enormous platform—the ability to say something to the world. First, how nervous would you be as you prepare? I've heard that public speaking is the number-one phobia, ahead of spiders, heights, the dark, and even death. People are more afraid to speak in front of an audience than they are to die. Based on this, I would imagine that there would be a healthy sense of fear as you approached your global speech.

In addition to managing your nerves, I imagine you would also spend a lot of time selecting the topic that you will speak on. The whole world will be able to listen. What would you want to say? Would you exhort people to world peace? Help the under-privileged? Save the environment? What would it be? This is your big chance. What kind of impression do you want to make? What kind of message do you want to deliver? If you were to speak to the world, I imagine you would be contemplating all these things.

What many of us do not realize is that the Internet allows us to communicate to the world. The entire world may not be listening, but if they wanted to, they could tune in and listen to you and me. Most of our messages on social media are public. Most of what we say can be seen not just by our friends who understand us and grant us a certain amount of grace, but also by perfect strangers who will be forming an opinion of us based only on what we say.

I'm pretty sure we all have the friend who constantly posts negative material. I have one friend in particular whom I have known since childhood. I think very highly of this person. He has been a very good friend for many years, but this person only posts the negative things that are happening in his life. I just want to ask him if he ever went back and read his posts. If he did, surely he would see the negative impression he is giving the world—a false and very skewed impression I would add. Facebook and Twitter are not replacements for therapy.

Communication, particularly on social media, is so convenient and easy that most of us tend to become lax in what we post. We have the ability to communicate with the world, and instead of making it a better place by exhorting some higher cause, we post pictures of our dinners, complain about our jobs, or list all the negative things that have happened to us in the last twenty-four hours. The ease of communication has lulled us into complacency and frivolity. Combine this with the fact that these messages and posts are preserved for perpetuity, and you can see why we all need to be careful about what we post. Digital is forever. You can't light a fire and burn your old posts. They can come back to haunt you.

Not to talk out of both sides of my mouth, but we also need to be prepared for the time when data that we absolutely need potentially goes away. I imagine most of us have heard of someone who has lost important files when her computer crashed. Losing data doesn't just happen when your computer breaks. Many cyber bad guys are holding data for ransom. A number of people and companies are infected with malicious programs that encrypt all their data, making that data useless. This encryption scrambles the data so that the owner can no longer read it. The bad guys will unscramble your data and leave you alone, but you have to pay them first.

The malicious programs used to do this are known as ransomware. A number of Ransomware programs exist, and they are very effective. In fact, at a 2015 security conference, a member of the FBI advised victims of these kinds of attacks to just pay the ransom.[1] The FBI has since retracted this advice. To avoid being stuck in this position, I recommend that everyone back up his or her data. Backing up your data not only gives you a second copy in case someone tries to extort you, it also comes in handy if your hard drive crashes or you accidently delete something. Copying your data solves multiple problems at once. There are a number of good solutions out there, and many of them now copy your data to servers on the Internet so you don't even have to buy any hardware, which makes setting them up and using them even easier.

In the area of personal cyber security, you should also think about passwords. Just to kick this off, let me say that "123456," "password," and "football" are not good passwords. I can go on and on about what a good, strong password looks like, but what people really struggle with is not trying to create good, strong passwords. Most people struggle with remembering all those passwords.

Fortunately, there is a solution for this problem. There are now programs called password managers that not only create strong passwords, but also keep them safe and secure. Password managers even remember your passwords for you. These tools allow you to use one very hard password to access a database that contains all your other user IDs, passwords, and even security questions. These programs encrypt all of your account information for all your accounts. The only way to unencrypt the information and use it is to enter a single password. This password is the only one you will have to remember going forward. Once you enter that single password, you will have access to all your other passwords. Some of the programs even type the passwords for you.

Think about the movie *Lord of the Rings*. In this movie, there is one ring to rule all the other rings. Imagine your one password is this ring. You can have one password to rule the plethora of passwords you have for your various emails accounts, shopping sites, banking sites, membership sites, social media sites, hobby sites, and the list goes on.

If you use one of these software packages and load all of your accounts and their passwords in it, you will be shocked at how many different accounts you actually have. I currently have over 100 different accounts, and I'm still adding new sites every other week. What I like best about these programs is that for most of the sites I have accounts on, I don't even know what the password is. I never even have to see it, let alone remember it. I can now have passwords like "'hm^XH7KvDqG,eK3++Npk?]JRUpE"—and not have to worry about remembering these passwords or even typing them. Password managers are a great way to not only help you keep track of all the login information for all your various accounts, but also to better secure those accounts with strong passwords. That's what we call a win-win!

If you want to take your password security to the next level, then you need to enable what is called "two-factor authentication." Two-factor authentication is basically the concept that you have to have two things before you can login. The first thing you need is something you know. That is your password. The second thing you need is something you have. This "thing" that you have is called the second factor. These second factors usually take on the form of a token with a number that changes every couple of seconds, a text message to your phone, or even a program that generates a number on your phone. In all these cases, you either need to have the token in your possession or have your phone in your possession. With two-factor authentication, you will need something you know (password) and something you have (token, phone, etc.) in order to log in.

The reason this second factor is so effective is because it is possible for someone to steal your password without you being aware of it. Think about it—if someone stole your password, would you know? It's not like your password doesn't work for you anymore. It's not like it is missing from your brain and you can no longer remember it after someone steals it. However, if someone stole your phone, how long would it take you to figure out that your phone is missing? When you need both a password and something you hold in your hand in order to login, it not only makes it harder for someone to login as you (they have to steal your password and your phone or token), but it also allows you to detect the theft much sooner if it does happen. For these reasons, two-factor authentication continues to be a very effective way of locking down accounts. Most major email providers and social media providers have some form of two-factor authentication available to their users. I highly recommend it.

In addition to shoring up your passwords, you will also want to practice good cyber hygiene. Definitely make sure that you have an antivirus program actively running and up to date on your computer. We talked a lot about random attacks and how automated defenses, like antivirus, are effective against them. I'm not saying antivirus will solve all your woes, but it will help to protect you. In chapter 9, we talked about the banking Trojan Zeus. Antivirus—although not a silver bullet—will still help protect your computer against these kinds of nasty programs.

You should also update the software on your computer on a regular basis. Most software now automatically checks for updates and lets you know when they are available. The problem is that you still have to install them. I'm a bit of an anal-retentive person, and I hate to see little red notifications on my screen, so I tend to install updates as soon as they come out—but I also realize that I'm not normal in this regard. Most of the time, those updates are not just to fix bugs, but to also close up security holes. Pay attention to updates and install them.

Finally, if you are thinking about protecting yourself and your personal data, consider purchasing a credit monitoring service. I consider this service the personal equivalent to corporate cyber insurance. Not only do these services help you identify attempts to steal your identity or gain access to your financial accounts, but many of them also offer some kind of insurance or guarantee and will help clean up any issues if they do arise. If you are the victim of identity theft, it would be really nice to have the assistance of an organization that has helped other people in the same situation before. Just being able to take advantage of their experience dealing with the aftermath of identity theft is valuable enough. If you have not already, I would encourage you to take a look at the options that are out there and select one that is right for you.

The tools and suggestions in this chapter are just a few tips that can help keep you safe. The reality of the situation is that the list of things you can or should do goes on and on. I tried to pick a couple heavy hitters that all of us can easily apply, but we have to constantly keep in mind that we are up against smart people who are always trying new ways to pull us in and take advantage of us. Vigilance on our part is critical.

NOTE

1. "FBI Advice on Ransomware? Just Pay The Ransom," October 22, 2015, accessed February 20, 2016, https://securityledger.com/2015/10/fbis-advice-on-cryptolocker-just-pay-the-ransom/.

Chapter 11

Cyber Security Is a People Problem, not a Technology Problem

When you have a hammer, everything starts to look like a nail. This popular saying contains a very poignant message for people concerned about cyber security. For years, the IT security industry has developed tools for addressing risk. These tools take many forms. We discussed some of them, such as the compliance frameworks developed around best practices. All of these tools have one thing in common: they were designed to manage risk.

Risk is a nebulous term that is often thought of as the probability of some event happening multiplied by the impact of that event happening. For example, in California the probability of an earthquake is bigger than it would be in, say, Florida. In each case the impact of an earthquake would be roughly the same. Contrast that with a hurricane. In this case the probability in California is low, while it is relatively high in Florida. Again, the impact is roughly the same. Using standard risk management (risk = probability × impact), we would say that California has a higher risk tied to earthquakes and Florida has a higher risk tied to hurricanes. Based on this analysis, California should spend more time addressing the risk of earthquakes and less on the risk of hurricanes and vice versa.

In the case of earthquakes and hurricanes, this risk analysis may seem like a complicated way of determining what is obvious. However, when you get into weighing more numerous and nuanced risks, such as whether a business should address risk associated with terrorist attacks or risk tied to natural disasters, this kind of modeling becomes helpful.

At a fundamental level, this risk analysis concept underlies most IT security programs. The reality is that regardless of how big your budget is or how many resources you have, nobody can address every risk. Because every organization is constrained, appropriately prioritizing which risks are addressed and which are not is a critical activity. The concept is to prudently assess and manage risk based on the probability of it happening and the impact when it does. For years this has worked very well . . . until now.

The problem comes when we realize that in the cyber security world, we are no longer managing risk. Risk is not personal. Risk is random. Risk has a probability that we can calculate and assess. Cyber security deals with people, and people turn risk models on their heads.

Consider the typical IT security prioritization dilemma of patching vulnerable systems. We discover new vulnerabilities in hardware and software almost every day. At first this may not seem like a problem, but think about the last time you had to install updates on your phone. The update probably took a couple of minutes and in some cases, your phone may have had to reboot, so you couldn't use it for a couple of minutes. Overall, it didn't disrupt you that much and was probably more annoying than anything else.

Take that experience and multiply it by 10,000 employees. In some companies, patching their systems means shutting down entire manufacturing facilities while the systems "reboot." Interdependencies also need to be tracked. In some cases, patches will actually break older programs. The last thing an IT team wants to do is apply a patch, and then find out that the patch broke an older system, and now that older system will not come back online. An IT team never wants to shut down the business by applying a patch. What you may experience as an inconvenience when you update your phone can turn into a major undertaking for companies.

With so many vulnerabilities and an often complicated process for closing them, companies need to prioritize which vulnerabilities they close right away, which they queue up to be addressed at a more convenient date, and which ones they will never get around to. To help with this process, most vulnerabilities are published with an industry-wide risk classification. This classification tells IT security professionals how big of a deal the vulnerability is. Organizations can then use these classifications to help prioritize their patching efforts.

As an IT security professional, you decide daily whether to focus your efforts on closing something that the industry says is high priority versus something the industry says is medium priority. Given a choice between fixing something classified as high priority versus fixing something classified as medium priority, which item would you work on? The high-priority item, of course!

The challenge in standard risk analysis lies here. Attackers know you will most likely fix the high-priority items first. If you are an attacker, and you can choose to use an exploit that is well known and high on the industry's list of things to close up versus an exploit this is not well known or low on the priority list, which would you choose? Attackers are incentivized to exploit what the IT security industry has deprioritized. Attackers target the weak spots that everyone has deprioritized, and this is exactly where standard risk management falls apart. In this situation, the items that the industry has determined as the most risky and should receive the most attention immediately drop off the attacker's list of items to exploit. The fact that the industry says it is high risk and needs attention means that attackers are less likely to exploit it.

The concept is kind of a chicken and egg argument. As the industry prioritizes a risk, security professionals start to fix it, and attackers start to move away from it. As this happens, in theory the risk should then go down, which means security professionals will stop working on it, which then means attackers will start exploiting it, and it goes on and on. With that said, the theory of risk modeling is sound, but when we use industry-wide information to set priorities for a specific company, a disconnect for that company often occurs.

My favorite experience around this risk comes from a CBS News Report. The report focused on the risks associated with multifunction printers. These printers are the big printer/copier/fax/scanner/slicer/dicer machines in most offices. The reporters explained that these devices store electronic copies of most documents they process, and when the devices are serviced or replaced, all that data stays on the device, even when the device leaves the company. Sensitive material then falls into the wrong hands.[1] After this report aired, my phone lit up with executives wanting to know how to prevent that from happening to our company.

My team and I spent several weeks trying to explain that we have a policy requiring all devices to be wiped clean of all data before they leave the property and that the policy is written into all our support contracts and audited on a regular basis. In other words, we had that covered. Regardless, we still spent weeks on it. Meanwhile, we had a group of hackers actively trying to break in through our external websites.

CBS News highlighted something that was high risk for most companies. Unfortunately, that risk didn't align with my reality at the moment. Rather than looking at printers, I should have been focusing on web servers, the same servers that were under attack. My particular risk did not match the industry's risk. If I had solely focused on what the industry said was the highest risk, I would have been the victim of a successful cyber breach.

Once again, we face the fact that shifting from IT security to cyber security means we are no longer fighting against risk, we are fighting against people. Because of this fundamental shift, most of the legacy IT security practices struggle.

Not only is the traditional approach to risk turned on its head, but traditional reliance on technology is also turned on its head. Once again, when you have a hammer, everything starts to look like a nail. Isn't it interesting that many technology companies advertise their technology as the way to end cyber attacks? Most IT security professionals, being technology people themselves, eat this up. They have been trained to use technology to solve problems. When a new problem presents itself, technology is always the answer, right?

When I started my job as a cyber security leader, I had a fairly strong technology background. Typical to my background, I viewed technology as the answer. My team and I had a voracious appetite for new technology. It was so bad that my information assurance leader became known as "The Tool Man" because he purchased and deployed so many different technologies to help us catch attackers. Of course, it didn't help that the technologies had names like Splunk, Snort, and Squil. I can still hear my leadership team now, "You want to spend how much deploying something called Snort?" It's a wonder my leadership team didn't think we were snorting something.

After a couple years of implementing the latest technology to address cyber attacks, we started to notice something. Three to six months after we implemented these technologies, the attackers would figure out a way around them and make them useless. In some cases, we even saw attackers use the technologies against us by sending us false alerts that distracted us from what they were really doing. The digital equivalent of a head fake!

When you think about technology, do you believe that anyone can create something that another person or team of people could never find a way around? Is it even possible? Remember the Enigma code machine and its 150 quintillion (remember, that's 150 with 18 zeros after it) different possible combinations? The code was supposed to be unbreakable. After much research and even the invention of some new technology, the Allies eventually broke it. Humans cannot create something that other humans can never figure out. Technologies that claim to solve the cyber security problem are either lying or ignorant of the problem they are trying to solve. I want to make sure you get that, so I am going to say it again. *Technologies that claim to solve the cyber security problem are either lying or ignorant of the problem they are trying to solve.* Either way, they certainly are not the final solution to your cyber problems.

If technology stops people from doing bad things, then why do we still need police officers, detectives, or even military? Just let the technologies stop the bad guys, solve all the crimes and prevent all the wars. This idea is a panacea that will not be realized. However, even if technology in and of itself is not the answer, that does not mean that technology cannot or should not be part of the answer. Just as in criminal justice, crime scene investigation, and military strategy, technology paired with people and process is what it takes to counteract other people with bad intentions.

In cyber security, there is a tendency to automate as many security processes as possible. Automation leads to people being taken out of the equation. Automation lowers costs and drives standardized results, but it also opens the door for attackers to exploit the automation. If the automation can be circumvented, then the defenses can be bypassed, and because technology delivers a standardized result, the defenses can be bypassed over and over. Strategy stops people. Reasoning stops people. People stop people. This doesn't mean, however, that there is no room for automation inside cyber security programs.

My organization did a study once to find out how average security analysts spend their time. We wanted to know how much time analysts spend on different activities as they searched and tried to identify attackers. We knew that most of an analyst's time was spent investigating alerts. These alerts are triggered whenever something strange or suspicious happens, kind of like a police officer investigating the report of a suspicious person hanging around a neighborhood. Sometimes those alerts are legitimate and sometimes they are not.

When an attacker causes an alert, it is called a "true positive." When an alert is caused by something other than an attacker, it is called a "false positive." Because the volume of these alerts is usually very high (thousands of alerts per analyst per day), the industry focuses on making the alerts better and making more of the alerts true positives, while driving down the number of false positives. In fact, many security technologies advertise how low their false positive rate is.

This concept of managing false positive rates is a bit of a balancing act. If too many false positives occur, then your analysts will be overwhelmed, and chances are they will end up missing the true positives in the noise. This is referred to as "alert fatigue" and is a common problem in the security industry. So many other alerts are firing that the legitimate one just gets lost. However, the pendulum can be swung too far in the other direction. By making your alerts so tight that they only fire when you absolutely know something is wrong, you end up missing attacks. This situation is known as a "false negative." The alerts were tuned to be so tight that they failed to fire when an attack took place. In an effort to block out all the noise, the attack was also blocked out.

Most of the time, attackers try to behave like a normal user as much as possible. Because of this behavior, often the only way to catch attackers is with an alert that fires even on normal activity. This balance between true and false positive alerts is kind of like walking a tightrope. You can fall off the right side or you can fall off the left side. Whether you are going off the right side or the left side, you are still falling, and that is not a good situation. You need to find your balance in order to walk the narrow path down the middle.

Let's look back at my organization's study about how analysts spend their time. When we understood true and false positives, we were able to appropriately tune our alerts in order to hit that sweet spot. When we found a good balance, we wanted to make sure our analysts were processing those alerts as effectively as possible.

As we examined our analysts' time, we expected to find that our analysts were spending most of their time deciding whether or not these alerts were true or false positives. At our first pass, this assumption was true. Analysts spent almost all of their time processing alerts. However, as we dug into what the analysts did to actually determine whether an alert was a true or false positive, we found something very interesting. Analysts spent almost 90% of their time gathering data so they could decide whether an alert is legitimate or not. We found that alerts contain small pieces of information around why the alert was triggered, but the analysts needed more context to decide whether it was a false positive or not.

Look again at the example of police officers investigating the report of a suspicious person. The officers receive the alert, check it out, and sure enough—a suspicious person is in the neighborhood. Is that person causing trouble or does he have a legitimate reason for being there? The police officers have to gather additional information to determine if the person is indeed causing trouble. They will most likely have to speak with the suspicious person to find out what he is doing and why he is there.

Several years ago, my wife and I noticed a car with one person in it sitting on the corner by our house. We live on a fairly quiet street with little traffic, so the car was very noticeable. After a couple hours the car was still there. We agreed that this was definitely suspicious. We called a couple of neighbors to see if they knew what was going on. The neighbors had also noticed the car and were equally suspicious. With nobody knowing why the car was there or what the person in it was doing, we alerted the police. An officer was sent out. He easily identified the car and the occupant. In other words, he was able to quickly verify why we had alerted him, but he still didn't know if there was a problem or not. After speaking with the man in the car, he learned that the man was a private investigator. All his paperwork checked out, and he was allowed to continue sitting in his car on the corner. The alert was a false positive.

Our analysts were doing the same type of investigation. They could quickly verify why the alerts had fired, but they needed additional information to determine whether something bad was happening or not. This information may include data in logs, information on a laptop, or routing data from an email—any number of things. To our surprise, analysts spent around 90% of their time gathering that information.

Looking at this, we quickly realized that the data gathered to figure out whether an alert was a true or false positive did not help us better protect ourselves. The data gathering was just overhead. Having a person look at alerts to decide whether these alerts are true or false positives is very valuable when you are trying to stop people. Your analysts can think on their feet; they can detect subtle changes in attackers' techniques. We needed them making those decisions, but we did not need them spending 90 percent of their time gathering data.

In other words, we do not want to automate the decision-making process around whether or not an attacker is in the environment. However, we do want to automate the gathering of the information that the analyst needs in order to make that decision. Judicious use of technology to increase the productivity of people in the cyber defense process is valuable. Using technology to replace people in your cyber defense process is not.

This flies in the face of what the security industry is saying. Just a quick look at marketing material out there shows technologies that protect against "99% of threats,"[2] systems that allow you to "detect and prevent threats, at any stage,"[3] as well as systems that "solve your security challenges."[4] All of these vendors sell very good technologies, but the technologies alone are not enough to keep companies safe. Unfortunately, numerous security organizations pin their hopes of success on these technologies.

Don't get me wrong—technology will help. However, technology is not the only thing you need. In fact, some technology that actively blocks attackers can inadvertently help attackers become successful in the long term. Recall the first principle of creating morphing defenses from chapter 9. Just like removing a bolt that you cannot see is much harder than removing one that you can, attackers find it much more difficult to avoid defenses that they cannot see and interact with than they do defenses that they can actively probe and prod.

Cyber security is not a technology problem; rather, it is a people problem, and you cannot use technology to solve people problems. Do not rely on technology as your only defense. Attackers will get around it every time. People need to defend against people. However, technology should and must be used to make your people as efficient as possible.

My learning from the analysis we did on analysts' time was not that we needed to reduce false positives or that we had to automate analysis. My learning was that we needed to automate the gathering of data so analysts can make informed decisions as quickly as possible. Automation does have a big play in cyber security. That play is not to replace the people on the front lines though. We must deploy it in a way that makes the people on the front lines more effective.

NOTES

1. "Digital Photocopiers Loaded With Secrets," last modified April 25, 2010, https://www.youtube.com/watch?v=6pIFUOav2xE.

2. "Cylance," accessed January 21, 2015, https://www.cylance.com/.

3. "CrowdStrike," accessed January 21, 2015, http://www.crowdstrike.com/.

4. "FireEye," accessed January 21, 2015, https://www.fireeye.com/.

Chapter 12

Facing Cyber Security Head On

Cyber security is on a lot of people's minds. Cyber security issues seem to have come out of nowhere—as though they are a new kind of evil that the world has never seen before, or a very complex set of challenges that are baffling the experts. Elements of truth are in these statements, but understanding cyber security in the broader context of the ongoing game of cat and mouse between attackers and defenders should give all of us a larger view of the problem.

We should not view the challenges in cyber security as something new or independent of what has been happening since the beginning of the human race. The technology has changed. We are no longer using sticks and stones to attack and defend; we are using advanced globally connected computers. This technology enables the battle between good and evil, between attacker and defender, to change more quickly than we have ever seen. This speed of change is new, but the fundamental problem of defending against people with malicious intent is not.

In this book, we looked through the development of the IT security industry and saw how a reliance on standards and an assumption that all attacks will impact a large number of computers has opened the door for the kind of targeted cyber attacks we are all reading about in the news. We also looked at how the Internet was created. We saw how this new technology that we call the Internet was architected without security in mind. Whether it was the original ARPANET that connected a small group of colleagues, allowing them to share data with people they trust, or the underlying protocols that were quickly put in place just to keep the Internet working as it grew, we saw that security is not fundamental to the Internet. Add in the fact that all technology becomes less expensive and easier to use over time—even malicious hacking technology—and hopefully you have a better appreciation for how cyber security has become such a big topic and concern.

Although cyber security is a technology-laden space, technology is not the answer to our problems. At the other end of every attack is a person—a person bent on stealing information—a person who uses technology, but a person nonetheless. Technology alone will not stop people. People still need to stop people. You also have to know what it is you are trying to protect and take appropriate measures to protect it.

Companies need to know whether they are protecting unique information or if they are protecting information that is more of a commodity and can be found in numerous other businesses. Understanding this distinction will help businesses see how high they need to raise the bar when it comes to cyber security. If your business is the only place in the world where a particular piece of information exists, then the bar will have to be pretty high in order to protect that information from a determined attacker.

We spoke earlier about creating dynamic defenses that present new challenges to the attackers every time they come back. This approach is very effective against cyber attacks. In this model, it is critical to have someone play the role of a cyber detective to direct the changes to these defenses. As your cyber detectives see the same attackers over and over, they should be able to observe how the attacks are executed and how they are changing, all the while drawing connections with attacks at other companies in order to determine what they can do to inflict the maximum pain on the attackers with the minimum impact on the business.

At the end of the day, if your business has critical data that it has to protect, having dynamic defenses and a cyber detective who can direct those defenses are the critical elements to protecting your organization against advanced cyber attacks. Implementing technology that does the same thing over and over will not give you the dynamism that you need. Using a process that looks at incidents in isolation will not give you the big picture view that you need to direct changes to your defenses.

Again, not every company has enough cyber risk to justify this level of defense, but your company may indeed need this level. Using both dynamic defenses and a cyber security detective is a proven approach that I have seen work in both large and small businesses. Using technology to automate the non-value added portions of this approach also makes it cost-effective and within reach of businesses of any size.

The Internet—and the technology that uses it—is changing the way businesses operate. This force is powerful and is opening a number of new opportunities. The future is exciting. However, with this opportunity comes risk. Cyber risk is not a technical problem. Cyber risk is a people problem.

Businesses need to understand this. Security leaders need to understand this. Addressing the kind of cyber attacks that we all hear about on the evening news requires a different mindset, a different view of the problem.

We are no longer stopping malicious programs. We are stopping people. We are no longer playing chess against a computer. We are playing chess against a person and that requires a different strategy and approach.

We have talked about some of the implications of this different view of the problem throughout this book. As simple and subtle as it sounds to shift from stopping bad programs to stopping bad people, the shift is seismic and should impact every element of your security program—whether it is viewing your attacker as your auditor or striking a balance between defenses that automatically block and tip your hand to attackers versus defenses that are invisible to the attacker and let you block at a point of your choosing. Either way, the implications of focusing on people and not programs are huge.

I was recently reminded of how different these two mindsets are and how the traditional security industry still has not made the shift. At a large IT security conference with hundreds of IT security leaders attending, many of them representing Fortune 500 companies, I had an opportunity to listen to a well-respected chief information security officer present on advanced attacks. This leader had been in the industry for many years, was considered a luminary, and carried a significant amount of influence. Because this leader was from an industry that I was not very familiar with, I was looking forward to the presentation and hopefully learning from his different perspective.

As the presentation progressed, I quickly realized this well-respected luminary in the IT security world was still solving the old-technology problems. He had not shifted his thinking in order to address a people problem. It was almost as if he was repeating what he had heard from technology vendors. He talked more about the technology that could now stop attacks and never once mentioned a strategy for dealing with attackers themselves. Unfortunately, not long after this presentation, this IT security leader's business made the national news because of an enormous breach.

Oddly enough, the vast majority of IT security teams inside businesses that experience a breach eventually end up referring to that breach as a life-changing event. Everything changes after you have been breached. Not just the amount of support you receive from the business, but even the way you understand and think about the problem changes. You realize that the old compliance approach to security is no longer effective. You realize that simply monitoring activity logs trying to find anomalies does not work. You realize that patching every security hole you can find is a losing battle. You realize that you are no longer stopping malicious computer programs, but that you are stopping malicious people. Unfortunately for so many, it takes a breach, a life-changing event, for them to make this transition.

Many times, I have had to inform IT security leaders that they have been the victims of a breach. When I do this, it is literally like watching someone go through the stages of mourning. First there is denial. The leader swears there is not an issue. There must be some kind of misunderstanding or the issue is not as big as I am making it out to be. It usually takes a lot of deep technical conversations in order to prove there is an issue, but once I do, they become angry. They are the victims of a crime, and many of them have told me that they feel like they were violated, as though their home has been broken into. For many security leaders, protecting their company is personal. At some point, the leader eventually accepts the situation he or she is in and usually begins to understand exactly what he or she is up against.

With that said, there is a saying in the industry that within a year of a company being notified of its first breach, the security leader will either have his or her budget doubled or a new job. There is a lot of truth in this saying, and I have seen both scenarios played out many times. It is the security leaders that are able to change the way they view their jobs and their understanding of the problem who survive. It is the leaders who remain stuck in their old way of thinking who do not.

The change in mindset does not stop with the security leader. Businesses also need to understand what they are trying to protect and then create a plan that matches their risk. We have painted a picture of how a company can protect highly sensitive and critical information with a dynamic defense that is directed by the equivalent of a cyber detective. This is the high end. Consider cyber security as a graduated scale that goes from doing nothing to creating a dynamic program like we outlined. Businesses need to assess their risk and determine where they need to fall on that scale. There is no one-sized-fits-all solution. Risk in the cyber world varies greatly from business to business, and as such, each business needs to determine for itself where it falls on that scale and then create a security program that puts it where it needs to be.

Whether we are talking about the change from IT security to cyber security, the difference between random attacks and targeted attacks, or various approaches to dealing with automated attacks verse attacks that are directed by people, at the end of the day, we are talking about the same thing. The very nature of protecting data, both business and personal, has changed. The IT security industry has spent decades creating approaches to security that no longer work against these new kinds of attacks.

The challenges facing those of us whose job it is to defend data now require us to continue practicing the old ways, while also developing new approaches. The old risks of random and automated attacks are still with us, and the old approaches of following best-practice lists—and automating as much as possible—still work for these risks. Unfortunately, a new set of risks now exists and these risks require new approaches.

Change is often hard for industries. After spending years developing successful models for defending against IT security issues, it is hard to know when those models need to change. The cyber security industry is at that point today. Gone are the days when the next killer technologies will save the day. Because of their human nature today's cyber security challenges are much more challenging and complicated. We are now dealing with a people problem, not a technology problem. People problems inherently defy automation.

Think about other kinds of people problems. Crime, for example, is a people problem. Have there been any effective automated solutions to crime? No. If there was a way to automatically prevent crime, I'm pretty sure it would be put in place very quickly. People problems need people-based solutions.

As time goes on, I believe the technology-focused industry will make room for very focused and capable managed service offerings. This new set of managed detection and response service providers will pair technology with the people and processes required to successfully defend against targeted cyber attacks. These providers will also be able to deliver these services at a level of scale that will make it hard for internal organizations to match not just the price but also the quality.

Consider what it takes to monitor a network for attacks on a constant and continuous basis (24 × 7 × 365). In order to provide shift coverage alone, a company would have to staff at least twelve full-time employees. This would not just allow analysts to take vacation and sick days, but would also provide multiple overlapping shifts to ensure adequate handoff time between analysts when an event is underway. In addition to having analysts constantly watching the environment, a company would also need specialists with incident response skills as well as specialists with threat intelligence skills. The analysts would constantly monitor the environment and identify the attacks. The incident response resources would spring into action to neutralize any attacks and ensure that the attackers are fully removed from the environment. Finally, threat intelligence resources would learn from the attack and direct all changes to the defenses as part of the morphing defenses we discussed earlier. With all these people and the varied skills sets across them, you quickly require a several million-dollar budget just to staff the organization, let alone purchase, install, and maintain the technology. That is of course assuming that you can find, recruit, hire, and retain all that talent, which in today's cyber security job market, is very hard to do.

Not only can these managed detection and response service providers deliver the various talent at a lower price point than most organizations could do internally, but they also have the added benefit of being able to see and experience more attacks than most companies. Because these providers will deliver their services and defend multiple companies across multiple industries, each of their customers will receive better protection. Recall the learning loop that we described previously where defenders should not just stop attacks, but learn from them and constantly adjust their defensive posture based on that learning? We discussed having the cyber equivalent of a detective directing the changes to the defenses. In the case of a managed detection and response service provider that has successfully implemented this learning process across all its customers, each customer will benefit from any attack at any other customer. This truly means that the sum of the defenses provided by the service provider can be greater than what each individual customer could provide for himself or herself.

All this said, the managed security service provider (MSSP) industry today still has a lot of maturing to do before they can step into this space. The vast majority of MSSPs simply view their job as monitoring log files and processing alerts. In a number of cases, MSSPs simply wait for a security technology in the customer's environment to send an alert that something strange has happened. Once that alert comes in, the MSSP will notify the customer. There is almost zero value being added. The alert is not investigated, the MSSP is not responding or cleaning anything up, the MSSP is not learning anything new, and the MSSP is certainly not changing the defenses of the customer. In essence, the MSSP could be replaced with an email. Alert comes in from a security device, an email goes out to the customers notifying them. Not only is there minimal, if any, investigation of the alert to validate it, the learning loop that we just discussed simply does not exist within the MSSP processes. This may sound like a simple thing to put in place, but it requires an entirely different mindset.

If this is the value that MSSPs are bringing to the market place, then why do so many companies employ them? Because most of the best-practice lists that are out there that companies are audited against require that someone monitors the log files. At the very least, MSSPs monitor log files. Simply put, MSSPs check the box and make auditors feel good and help you pass audits.

There used to be a saying in the IT industry that nobody got fired for buying IBM. That was even after IBM fell behind on the technology curve and was providing less value than its competitors. The security industry is in the same place. Nobody gets fired for hiring a big name MSSP.

I recently spoke to one large financial institution that hired one of the big MSSP firms to monitor its logs. I asked them what kind of value they were seeing from the service they purchased. In a moment of blunt honesty, the IT security leader at the company told me that the value they get is a check mark from their auditor. The auditor sees that firm at other businesses and is comfortable with the firm. When asked about the quality of the alerts or how well the MSSP actually protected their environment, the IT security leader told me that his team doesn't even look at the alerts the MSSP sends over. He told me they only pay the MSSP because the auditor will approve. His internal team does all the real work protecting the business. This business model is not sustainable for MSSPs.

Service providers can no longer focus only on the alert that is in front of them at the moment. They need to be able to connect dots in order to identify the multiple threads that make up an attack and pull them all together. Where today's MSSPs deliver the digital equivalent of beat cops, tomorrows managed detection and response service providers will deliver those beat cops in addition to the detectives that can see more broadly and direct constant changes to defense. Managed detection and response service providers will have to move from simply monitoring log files to playing an active role in the definition and maintenance of each customer's defensive posture. In a technology-laden industry, this type of service is sorely lacking today.

As you can see, cyber security is not a plug-and-play or paint-by-numbers exercise. Following a set of predefined best practices will no longer cut it. Stopping attackers requires strategy to out think the adversary. Stopping attackers requires creativity to find new and innovative ways to inflict damage to the attacker without impacting the business. Stopping attackers requires agility and technology that doesn't stop attacks, but that enables you to realize your unique strategy and implement your innovative ideas.

While IT security was a stodgy, behind-the-scenes exercise of "doing the right things," cyber security is an exciting, dynamic environment that doesn't have any rules—an environment where success is defined by whether you win or lose. It is you, your team, your technology, and your intellect against an adversary. The stakes are high, but with the right approach, defenders have and will continue to succeed.

Additional Cyber Security Resources

CYBER STRATEGY AND TRADECRAFT

"Intelligence-Driven Computer Network Defense Informed by Analysis of Adversary Campaigns and Intrusion Kill Chains," http://www.lockheedmartin.com/content/dam/lockheed/data/corporate/documents/LM-White-Paper-Intel-Driven-Defense.pdf.

"Ten Strategies of a World-Class Cybersecurity Operations Center," https://www.mitre.org/sites/default/files/publications/pr-13-1028-mitre-10-strategies-cyber-ops-center.pdf.

Cole, Eric. *Advanced Persistent Threat: Understanding the Danger and How to Protect Your Organization.* Syngress, 2012.

Crumpton, Henry A. *The Art of Intelligence: Lessons from a Life in the CIA's Clandestine Service.* New York: Penguin Books, 2013.

Wrightson, Tyler. *Advanced Persistent Threat Hacking: The Art and Science of Hacking Any Organization.* McGraw-Hill Education, 2014.

"Operational Levels of Cyber Intelligence," xhttp://www.insaonline.org/i/d/a/Resources/CyberIntel_WP.aspx.

CYBER HISTORY

Healey, Jason. *A Fierce Domain: Conflict in Cyberspace, 1986 to 2012.* Cyber Conflict Studies Association, 2013.

Stoll, Cliff. *The Cuckoo's Egg: Tracking a Spy Through the Maze of Computer Espionage.* Pocket Books, 2005.

Corera, Gordon. *Intercept: The Secret History of Computers and Spies.* Weidenfeld & Nicolson, 2015.

TECHNOLOGY

Carvey, Harlan. *Windows Forensic Analysis Toolkit, Fourth Edition: Advanced Analysis Techniques for Windows 8.* Syngress, 2014.

Ligh, Michael, et al. *Malware Analyst's Cookbook and DVD: Tools and Techniques for Fighting Malicious Code.* Wiley, 2010.

Sikorski, Michael, and Andrew Honig, *Practical Malware Analysis: The Hands-On Guide to Dissecting Malicious Software.* No Starch Press, 2012.

Bejtlich, Richard. *The Practice of Network Security Monitoring.* No Starch Press, 2013.

Murdoch, Don. *Blue Team Handbook: Incident Response Edition: A condensed field guide for the Cyber Security Incident Responder.* CreateSpace Independent Publishing Platform, 2014.

Index

About the Author

Brian Minick is the cofounder and CEO of Morphick Cyber Security. At Morphick, Brian delivers managed detection and response services that enable Morphick customers to address emerging cyber risks. Brian brings over fifteen years of diverse information technology and cybersecurity leadership and experience. Before founding Morphick, he held the title of Chief Information Security Officer at General Electric's Aviation, Energy, and Transportation businesses where he was responsible for developing and implementing advanced cybersecurity strategies. His approach was integral to protecting the strategy, growth, and resources of a multi-billion-dollar, worldwide corporation.

His experience at GE afforded him the opportunity to study cybersecurity from multiple perspectives including prevention, detection, and breech remediation. In addition, he led efforts to implement protocols for mitigating advanced persistent threats for an organization of more than 135,000 employees worldwide. Brian oversaw the implementation of security strategies that required multi-industry collaboration, regulatory compliance planning, and management of infrastructure obsolescence.

Brian lives in Cincinnati, Ohio, with his wife and three daughters. When not preventing cyber attacks, Brian has a leadership position at his church, New Life Chapel. He also enjoys playing with his daughters, running, and reading.